JUST IN CASE

JUST IN CASE

A Manual of Home Preparedness

BARBARA G. SALSBURY

ILLUSTRATIONS BY LARRY P. SALSBURY

Salt Lake City, Utah

Library of Congress Catalog Card Number: 75-13977

ISBN 0-88494-280-5

7th Printing, 1980

Lithographed in the United States of America
PUBLISHERS PRESS
Salt Lake City, Utah

Contents

Tables

Preface

This is a book on being prepared. Its purpose is to provide the know-how (and perhaps some incentives) for setting up a realistic, organized program of home preparedness, whether for an individual or a family, suitable to the circumstances and environment. It will help you set and achieve preparedness goals which will induce a feeling of security about meeting future emergencies.

If, as has been said, experience is a good teacher, the author has been taught well. Let me share with you what I have come to call the "Salsbury Saga." It's true.

In December 1967, two days after Christmas, a heart attack put me in the hospital. The following February I was returned to the hospital with a blood clot in my leg. In March our young son tried to telescope (or pass) his own bowel. He was rushed to the hospital and given several hours to live. He lived.

Spring came, and that meant baseball. My husband loves baseball. One April evening, playing in the outfield, he stretched to catch that long fly when his cleats caught in the mud and he ripped the hamstring muscles in his right thigh. During his hospital stay he caught double pneumonia. In June that year I was again in the hospital, this time for triple major surgery.

July arrived and we decided to drive to northern California, where my husband obtained a job. Upon returning home, he packed and labeled everything, and I was to stay until the house was sold. Our house was broken into the night that Larry left, and many of the packaged boxes were stolen.

The following week I discovered we had been caught in a land fraud and really did not own the house after all. After discussing it with several lawyers, I called Larry and said, "Come and get me!" Larry arrived, but the rental truck, which had been half-loaded by friends, was too small. At this point I could drive, but not pull a trailer. Faced with the question, Shall we take the furniture or the food and provisions? we decided to take the provisions. Two weeks later when we returned for our furniture it was gone. We went back to northern California empty-handed.

All of this happened in just one year! But these and other similar experiences taught us much about preparedness. Through that whole unbelievable year we were able to take care of ourselves. Not once did we have to choose between proper medical care and feeding the family. Not once did I have to worry how my tiny children would be fed while my own hold on life was so precarious. We didn't go under. We survived!

This book has been written to share with others some of our experiences, our trials and errors and successes. I sincerely hope it will impress you with the importance of preparing to take care of your needs in an emergency and move you to act accordingly.

I realize that the home preparedness field is growing and that new methods, new knowledge and new items are being developed almost daily. I also recognize that in this book I have only dented the surface of the helps available at time of writing. Nevertheless, the methods, ideas and thoughts given here will basically do the job. I offer them so that you might be prepared . . . just in case.

Part I
FOODS

1

Nutrition and Common Sense

One of today's most widely discussed and controversial subjects is nutrition. The word has almost become a "byword" in any consideration of food products. Yet, among average people, confusion reigns about this subject.

Nutrition is important. Yet most of us know very little about the nutritional value of available food products. We are tossed to and fro, and sometimes imposed upon. Rather than be swept with the tide, we need to have a basic understanding of nutrition, sprinkled with a lot of common sense. For instance, we must not emphasize one factor at the expense of another. Because the word is so well known, vitamins tend to receive this emphasis. But although they are "necessary for good nutrition, they are no more important than are proteins, fats or carbohydrates. We need them all in their proper proportion."[1]

Good health and energy from the food you eat are goals you should have for yourself and your family. If an emergency situation arises, these goals should remain the same — to be able to maintain good health under all conditions. By preparing *now,* you will be able to have more than a mere existence *then.*

Basic-Four Food Guide

The essence of the nutritionally well-balanced diet has been broken down for us in what nutritionists call the Basic Four.

> The basic four food groups should be interpreted as just what the name (basic) implies — a foundation upon which to

[1]Elna Miller, *Facts About Food and Nutrition* (Logan, Utah: Utah State University Extension Service, 1973), p. 13.

build good meals. Each food group makes special contributions to our bodies. Foods from all four groups work together to supply energy and nutrients necessary for health and growth.[2]

Good nutrition requires that you follow the Basic Four food guide in your meal planning.

Milk and Dairy Products Group

Milk is our leading source of calcium for bones and teeth and other tissues. It also gives high quality protein, riboflavin, vitamin A, and other nutrients. Amount needed daily: Children — 3-4 cups; teenagers — 4 or more cups; adults — 2 or more cups.

Meat Group

This group gives our main source of protein, which we need for growth and repair of body tissues, muscles, organs, blood, skin and hair. These also have iron, thiamine, riboflavin, and niacin. Amount needed daily: 2 or more servings — beef, veal, pork, lamb, poultry, fish, eggs; alternates — dry beans, dry peas, nuts.

Vegetable-Fruit Group

These are the most valuable for minerals and vitamins. This plan supplies most of the vitamin C needed for healthy gums and other body tissues. It gives over half of the vitamin A each day for growth, normal vision, and healthy condition of skin and other body surfaces. Amount needed daily: 4 or more servings — citrus or other fruit or vegetable important for vitamin C; a dark green or deep yellow vegetable for vitamin A, at least every other day; other fruits and vegetables including potatoes.

Bread-Cereal Group

This group supplies iron, several B vitamins and food energy as well as some protein. Amounts needed daily: 4 or more servings — whole grain, enriched, or restored.[3]

Keep these basic ideas in mind as you provide food for storage. They will be a great help.

Getting the Necessary Nutrition

How will you obtain the proper food value from foods you plan to store? For example, a ton of wheat in the basement is

---

[2]"Proper Nutrition," *Relief Society Magazine* (Salt Lake City: The Church of Jesus Christ of Latter-day Saints, 1970), p. 854.

[3]Miller, *Facts About Food*, p. 4.

wonderful, so long as you know the food value of wheat and the proper ways to prepare it.

The accompanying charts contain information to help you evaluate foods according to the Basic Four food groups and maintain a proper balance in the foods you choose.

ESSENTIAL FOOD NUTRIENTS — FUNCTIONS AND SOURCES

Nutrient	Functions	Food Source Information	Food Group
Protein	To repair and build body tissue.	Complete proteins — provide life and growth; derived from animal foods, meat, milk, eggs, etc.	2, 1
	Source of amino acids used as building blocks by the body.	Incomplete proteins — maintain life but do not provide for growth; derived from vegetable foods, grains, etc.	4
Minerals			
Calcium	To develop strong bones and teeth. To regulate activity of nerves and muscles.	Most readily available form comes from milk and milk products.	1
Phosphorus	To insure normal blood clotting. Frequently associated with calcium in the body; needed for bones, teeth; necessary to life of every cell, and nerve tissue.	Widely distributed in foods.	2, 4, 1
Iron	To build red blood cells and is necessary to every body cell.	Small amounts needed; can be obtained by sufficient protein foods — egg yolk, dried fruits and enriched grain foods.	2, 3, 4
Vitamins			
A	For growth, normal eye function and health, skin and mucus membranes.	Bright yellow color in foods; fat soluble; liver, egg yolk, fish liver oil, cream, evaporated milk.	1, 2, 3

Nutrient	Functions	Food Source Information	Food Group
B	Individual vitamins regulate processes in digestion; help maintain normal functions of muscles, nerves, heart, blood; essential to growth.	Widely distributed in foods, water soluble; variety meats, pork, ham, enriched breads and cereals, wheat germ and yeast.	1, 2, 3, 4
C	To maintain healthy bones, gums, cartilage, connective tissue; helps build blood cells.	Not stored in the body; easily destroyed by heat, light and air; water soluble; provided by fresh raw citrus fruits and some raw vegetables.	3
D	Essential for proper use of calcium and phosphorus in building bones and teeth.	Provided by vitamin D fortification of certain foods; fat soluble; also fish liver oils and sunshine.	1
Carbohydrates	Chief sources of calories; burned as fuel in the body for energy and heat; also provides roughage.	Cellulose in vegetables, fruits, cereals, nuts.	3, 4
Fats	Burned by the body for heat and energy; are carriers of fat-soluble vitamins A and D.	Some fats in foods (meat, milk, eggs, cereal); other fats purchased as fats (lard, butter, margarine, oil).	1, 2, 3, 4, other foods

Eat What You Store, Store What You Eat

If you follow the wise advice that has been given, you will use the foods you are storing, they will become a part of your accustomed diet, and you will acquire a taste for them. Do not buy anything you don't like and won't eat now, hoping for conversion later because "it's good for you." It is of limited practical value to have volumes of food lying unused or simply stored for some future emergency. These foods should be worked into the diet and used. It is important to have an excess to rely on, but it is equally important to be aware of the characteristics of various foods; to know how they need to be prepared; to know if a sudden overabundant use of a specific food will cause dietary problems.

A time of crisis is not the time to learn how to "concoct a new conglomeration."

Some people claim that it doesn't matter what kind of foods you have stored, that "if you get hungry enough you'll eat anything." That is a totally wrong concept.

> Dr. Norman Wright of the British Food Ministry, after experiencing conditions following the Second World War, said, "A sudden emergency is no time for introducing untried novelties." He indicated that people were more likely to reject unfamiliar or distasteful foods during times of stress than they would even under normal conditions. When we are frightened, upset and insecure, do we not return to things with which we are acquainted? Dr. E. M. Mrak, a member of the Division of Food Technology of the University of California, states that "it is a falsehood that a person will eat anything if he is hungry. There is abundant evidence to show that people have gone without food and have even lost weight rather than eat foods with which they are not acquainted."[4]

A Word About Food Fads

Here is one nutritionist's comments on food fads:

> Many different food fads are being promoted by many different people. The lay person, who has not been trained in recognized schools of nutrition, is bewildered by these conflicting ideas.
>
> Most of the food faddists know a little about sound nutrition facts, but they know a whole lot more about how to sell their spectacular ideas and expensive products.
>
> The United States has become a fertile field for the salesmen of the so called "health foods" and "miracle supplements." There is also an endless supply of other products, most of which add little to good nutrition but add much needless cost to your family's food supply.
>
> There are dangers from food fads and fallacies. Here are four of them:
>
> 1. Essential nutrients may be lacking in the fad diets. Malnutrition can result.
> 2. Children who are victims of a fad diet may not get the nutrients they need for proper growth and development.

---

[4]Merritt H. Egan, M.D., *Home Storage Advice Prepared and Compiled*, February 1959.

3. Medical attention to a serious ailment often is delayed while a food quack or faddist attempts to treat the condition.
4. Faddish foods and treatments are always expensive.

Feed your family, including yourself, sensibly, intelligently, and economically, but by all means *feed them food.*[5]

The problem of course is to distinguish between the approach of the true "faddist" and the legitimate search for a more natural, healthful way of living. Many who would scorn the faddist approach are genuinely concerned about the possible effects of such food-industry methods as chemical additives to food — colorings, flavorings, and preservatives; "refinement" of foods; and heat used in food processing. News reports on nutritional matters make us all aware that our food industry is far from perfect and that new knowledge can now and then surprise even the "orthodox" nutritionist. Clearly a person can hardly engage in a more vital search for knowledge than in this field. Equally clearly, dogmatism is an inappropriate attitude.

So far, the overall concept of the Basic Four seems not to be seriously challenged, so we may take this as the safe way. A proper diet based on this concept should provide the nutrients necessary for good health. Even then there will be unexplained exceptions, situations in which a doctor's aid may be necessary.

---

[5]Miller, *Facts About Food,* pp. 3, 15.

2
Getting Started

Initial Considerations

These are some important points to consider before you start buying a lot of food. Several things will determine the kind of preparedness program you should undertake.

First, realize that your pantry, your larder, your reserve, or whatever, is exactly that. It's yours. It's individual. There is no set, perfect, ideal pattern for everyone to follow. Your needs will vary according to your individual or family taste, situation, and circumstances. Here are some conditions that will affect your decisions.

— Your likes and dislikes of various foods.
— Special diet problems, such as sugar problems or allergies.
— The size of the family.
— The degree of physical activity. (Occupations of family members. How much strenuous activity is involved in their daily life?)
— Local climate, humidity, etc. (Affects your methods for keeping foods.)

If you have a family, plan adequately. Don't shortchange anyone. It would be far better to have a little extra than not to have quite enough.

How to Begin

After the decision has been made — yes, you definitely want a reserve put away — an ugly monster always rears its head.

It may be the "it's too big, I can't do it" monster; or the "it's too much work, it's impossible" monster; or the biggest demon of all, "I can't afford it."

Ignore such monsters. There are several ways to accomplish the project even without going into debt. It takes time, persistence, patience, effort and usually some money, but it *can* be done. Try some of the following suggested methods, adapting and combining them to suit your circumstances.

— Plant a home garden. Plant enough to eat now and preserve the surplus.

— Do home canning, home freezing, home drying. Economical when the produce is purchased in bulk at the peak of the season.

— Cut meat costs. Take advantage of hunting and fishing seasons, seasonal meat sales, etc.

— Buy staples in large amounts. Every other month or so, take the entire month's food budget and buy a case each of various items that you frequently use (tomato sauce, green beans, etc.). Feed the family from these items. This procedure will quickly build stocks.

— Overbuy. Every shopping day buy three or four packages or cans of shopping-list items instead of just one.

— Allot storage money. Allot yourself a set amount every shopping day strictly for storage foods.

— Shop the ads. Allow yourself a set amount each week or payday to take advantage of promotional sales at the local markets.

— Plan menus by inventory — another approach to storage-building. First inventory exactly what food and supplies you have, and write it down. Plan your menus from this list plus two or three storage items you must now buy to balance out and extend the foods on hand. For example, if you have tuna, noodles and other items for main dishes, you may want to obtain some fruits and vegetables for balance.

Always plan your food supply for variety, so that in an emergency you could prepare nutritionally balanced meals. It would be wise not to buy all fruit, all vegetables, or all of any one item at first. Obtain basic foods first; then, if you desire, expand to the luxuries.

Upon obtaining the kinds of items you want to store, work them into your everyday meals. Habitually work more and more of your grocery budget into storage items. Continue including storage-type foods in all of your meals. You will rapidly reach the point where you have enough items available from which to prepare meals; then food money can be used for additional items to whatever extent you wish.

This method serves two important purposes. It provides the means to buy your storage food. It also gives you experience. You learn to measure, season, cook and eat whatever types of food you have stored.

Whichever approach you use, don't become discouraged. Building a food reserve takes time. But this time and experience is priceless. Laugh together at your mistakes, then try again while you still have the opportunity to learn and try.

3
Key to Food Storage

No food can be stored without eventual deterioration. Proper storage conditions will minimize deterioration, however, and keep food safe from contamination and insect damage as well as from changes in flavor and appearance. Providing the right conditions takes knowledge and planning. Keeping food *dark, dry* and *cool* is the key to proper storage, the only exception being the root cellar (chapter 13).

Among the factors that can affect the shelf life and quality of food to be stored for an extended time are temperature, moisture, air, insects, and rodents. The real "villains" are heat, moisture, and air, for the right (or wrong!) combination of these three factors encourages bacteria and molds to grow, causing spoilage in low-moisture foods such as dried fruits. Flavor and appearance changes are also often caused by too much heat. The higher the temperatures in the food storage area, the more rapid the deterioration.

Given heat, air, and moisture, then, weevils are likely to infest grains and dried products. But please *do not* use a commercial fumigant. A product that kills insects can be extremely dangerous to humans, not only because we then consume the residue left on the food, but because toxic fumes are deadly in enclosed surroundings. Do not attempt to use unfamiliar chemicals.

Tale of the Weevil and the Bay Leaf

Once when I taught a home storage class, the discussion turned to some of the "old wives' tales" connected with food storage.

One such tale says that if bay leaves are put in wheat, grain or other stored items, the weevil will either be killed or kept out completely. When I stated that this information was incorrect, the reaction was one of great skepticism. Nearly all of the students had heard and believed this tale.

Research failed to disclose information indicating that the bay leaf had any toxic properties that would repel or kill insects. I then decided to resolve the question for myself.

I called the local feed and grain store, told them who I was and asked if I might come down and catch some weevil. After a few seconds of silence, the man on the phone responded, "You're kidding!" When I explained why I wanted them, he finally admitted they *might* have one or two. I entered the store with a bottle in hand and was escorted rapidly to the back — behind the large stacks of grain. By moving sacks and climbing over, around, and under, I managed to catch six black beetles.

Here's how the experiment was set up. On October 25, 1973, two beetles were put in jar number 1 (the control) with no bay leaves; two beetles and one large new bay leaf (crushed so that any "fumes" would be more potent) were put in jar number 2; two beetles and one old bay leaf (crushed) were put in jar number 3. A small amount of barley and of wheat flour were placed in each jar. Lids were tightly placed on the jars, which were dated and labeled.

At time of writing, nine months later, there are approximately sixty beetles in *each* jar. The weevils show no aversion to either old or new bay leaves. They ignore the leaves, crawling over, around and under them. This beetle has been identified as the sawtooth grain weevil *(oryzaephilus surinamensis* [*linnaeus*]).

Thinking that six weevils might not be sufficient proof, I set out to find more. It is difficult to come right out and ask a friend if she has weevils and still keep her as a friend. But such a friend was found, and on October 26, 1973, I brought fifty-two weevils home. These were a different kind from those previously used and were in the larvae stage.

Larry now designed a sophisticated "lab" of seven test tubes connected with glass tubing in such a way that the weevils had access to all seven tubes at all times. This "lab" was laid on a towel-covered cookie sheet, so that the tubes need not be disturbed while we observed the weevil. Numbering from left to right, the following items were put in the tubes:

Beetle "Laboratory"

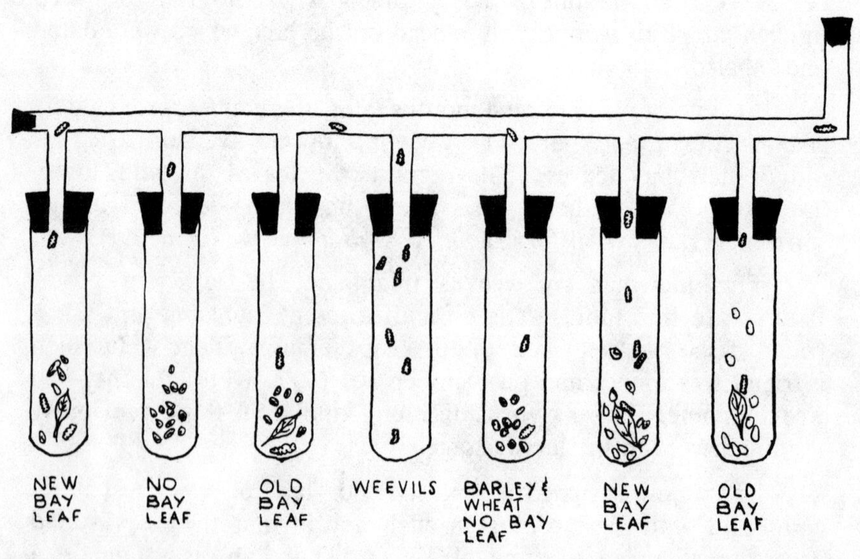

Weevil Larvae "Laboratory"

1. Large pieces of new bay leaf, whole wheat flour, wheat.
2. Whole wheat flour, wheat, no bay leaf.
3. Large pieces of old bay leaf, whole wheat flour, wheat.
4. Fifty-two wiggling weevils.
5. Barley, whole wheat flour, no bay leaf.
6. Large pieces of new bay leaf, whole wheat flour, wheat.
7. Large pieces of old bay leaf, whole wheat flour, wheat.

The weevils placed in the center tube varied in size from newly hatched and minute to good-sized and fuzzy. In order to obtain food, they had to go from the center tube into another tube of their choice. A log was kept of their progress and reactions.

After six months there were tiny weevils and large, fuzzy weevils progressing from tube to tube, crawling on or burrowing under all of the bay leaves. So far as we can tell, none have died. They continue to molt their skins, grow larger, and turn into black beetles. This weevil has now been identified as the carpet beetle *(anthrenus scrophulariae).*

My son decided this would be a good project for his science fair. He took twenty-seven larvae from the weevil "incubator." We kept a large jar of weevil-infested macaroni and flour, with a very tight lid, in case the first experiment failed. Using five test tubes instead of seven, he entitled his experiment, "Do Bay Leaves Help?" Since his results were exactly the same as ours, his conclusion too was that bay leaves do not protect food from weevils.

How to Deal with Insects

What do you need to know and *do,* in order to prevent infestation of your food supply by weevils?

First, realize that you cannot keep weevil out of your grains, flour and cereal products. The eggs, if not the insects, are already there. The weevil comes home with you in packages from the grocery store and in sacks of grain and flour from other food outlets. Your goal should be to prevent the weevil from growing.

In a storage area with a constant temperature between 35° and 50° F. weevil will probably remain dormant; but if heat and moisture are a problem, dry ice may be used to prepare cereal grains and other products for storage. This method is described in this chapter.

If dried foods such as grains, beans, cereals and nuts have

become infested with insects, spread the food evenly on a cookie sheet and place it in a 140° F. oven for one-half hour. This much heat will kill the insects, but grains treated in this way will not sprout.

To salvage insect-infested dried fruit, plunge it into boiling water. Dry the fruit thoroughly at a low oven temperature before storing it again.

Minimize infestation of bulk foods by storing them in airtight containers of metal, plastic or glass.

Attitude

When it comes to bugs and foods, attitude is important. It is hard to convince most homemakers that there is nothing wrong with a weevil, but if an emergency should arise it may not take you long to decide who is going to survive — you or the weevil. In some situations, you may not be able to throw out infested foods and buy more.

Unlike the cockroach or mouse, the weevil is a clean bug. Raised in whole wheat, it turns brown like the kernel; in whole wheat flour, it is beige; in white flour it is white. You can sift weevils out and eat the food anyway; or you can ignore them altogether. A little weevil has never hurt anyone (that we know of!).

Containers

Because it can be made airtight, a metal can with a lid that seals is the best container in which to store food products. The best kind of lid, called a "paint lip lid," is the tight-fitting kind such as is used on a paint can. The contents will be kept away from light. A five-gallon container is a convenient size to work with, because it can be moved easily and either the round or square style stacks well.

Many round metal cans being used for storage have lids that fit over the top and are not airtight. For extended storage of grain, beans, etc., the lid on this type of can should be sealed. (See below for instructions on sealing large garbage cans.) This type of container often can be purchased inexpensively from a large restaurant or bakery. Wash such cans well and allow them to dry thoroughly before using them as storage containers.

New garbage cans also are being used to store grain, beans, etc. Since this kind of can does not have an airtight lid, seal it with

EFFECTS OF HEAT AND COLD ON COMMON PANTRY INSECTS

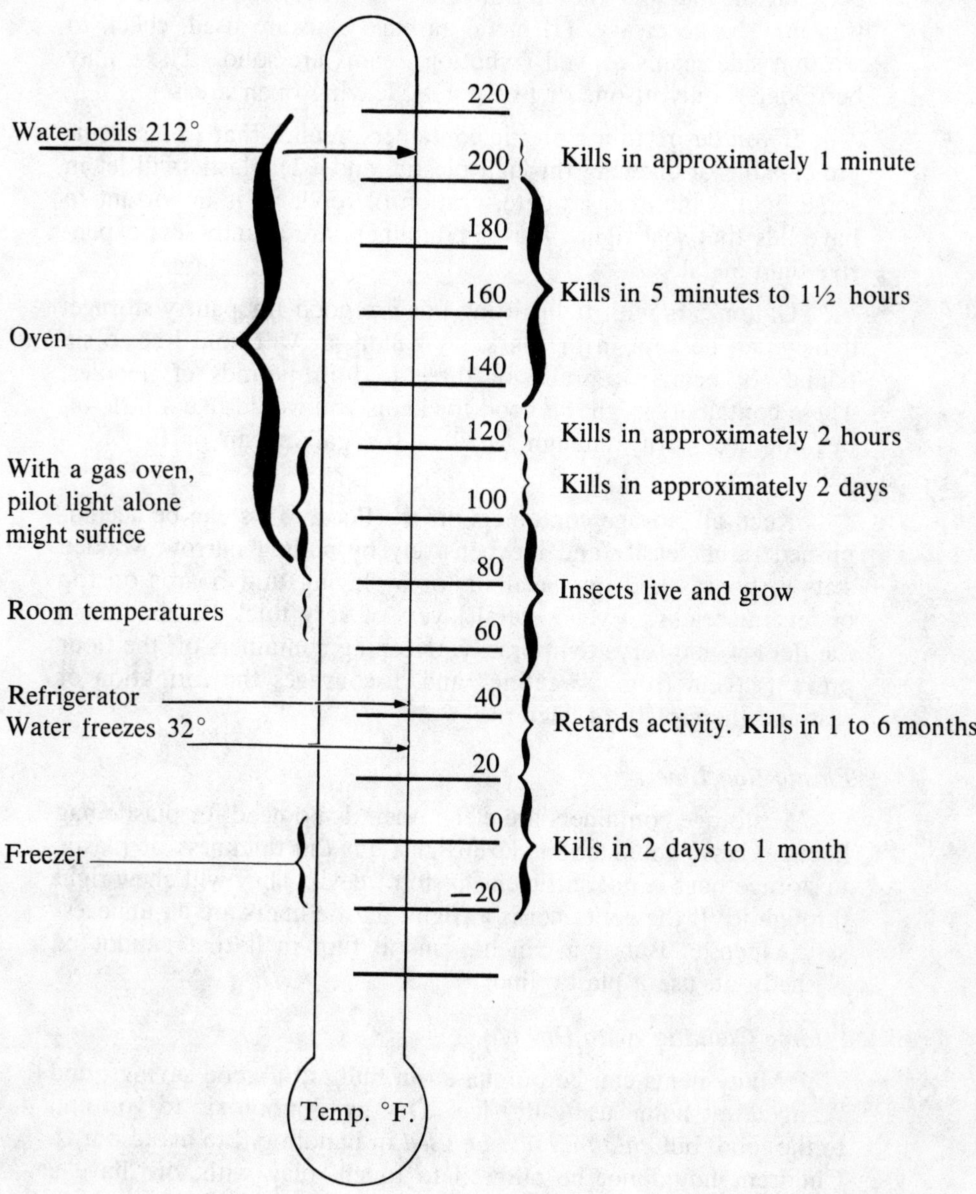

USDA and the University of North Dakota Extension Service.

several layers of heavy insulating tape, or use paraffin wax. Wax may be less expensive, but it takes more effort to be sure the crack between the lid and the can is filled in. Several applications of wax may be necessary. (If metal garbage cans are used, check to see that side seams as well as bottom seams are solid. These may be soldered only at one or two points, leaving open areas.)

If you desire to use plastic containers, realize that (1) rodents are capable of chewing through plastic, and (2) plastic will let in some light, which causes deterioration of food. It is important to have lids that seal tight. Plastic containers are usually less expensive than metal.

Gallon jars with tight-fitting lids are good for pantry storage, if they can be kept in the dark. A gallon jar will hold five to six pounds of beans or grain, or three to four pounds of noodles. These containers might be good for items you would like a little of, but not necessarily enough to fill a five-gallon can or fifty-five-gallon drum.

Keep all storage containers off the floor. This can be accomplished rather easily and inexpensively by nailing narrow wooden slats to boards laid on the floor, or by laying thin boards on top of several bricks. Even several layers of very thick cardboard on the floor would serve temporarily. Keeping containers off the floor prevents them from "sweating" and discourages the formation of rust and the growth of bacteria.

Plastic Bag Liners

If storage containers are clean, there is no need for plastic bag liners. Cans should be made airtight, but the thickness of plastic in storage bags is not sufficient to stop weevil. They will chew right through it. If the container is airtight, plastic liners are an unnecessary expense. But if a can has an oil film in it that cannot be washed out, use a plastic liner.

"Home Canning" with Dry Ice

Many items can be purchased in bulk, at a good savings, and "canned" at home using dry ice. Dry ice is nontoxic to you and to the food, but *caution must be used* in handling it to avoid burns. Children should not be allowed to touch, play with, or "have a bite of" dry ice. You should wear gloves or use folded paper to handle it.

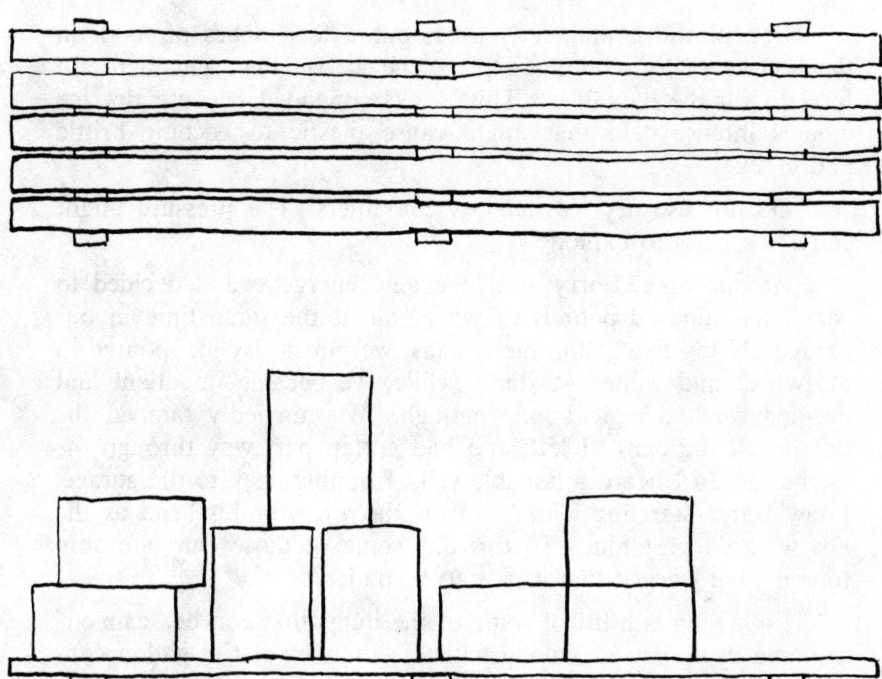

Floor racks, used for keeping storage containers off the floor

Dry ice is solidified carbon dioxide. At room temperature dry ice placed in the bottom of a container filled with grain or other storage food disappears, leaving carbon dioxide gas which forces the air and moisture out of the container. Insects will suffocate in this inert atmosphere, but this procedure does *not* kill any insect eggs in the food. After processing food this way, air and moisture can get back in if the can is opened many times, and the eggs could then hatch. If a proper container is used and it is kept sealed, the inert atmosphere will remain.

It takes eight ounces of dry ice per hundred pounds of grain; two ounces (approximately two square inches) per five-gallon can. Put the ice, crushed or in a chunk, on the bottom of the container, then pour the food on top of the ice. For a five-gallon can, wait thirty minutes before securing the lid. Allow two hours or more for a fifty-five-gallon drum. If the can sides bulge, "pop" the lid and wait a while longer.

If a plastic container is used, put a few inches of food on the bottom of the container, place the ice on this, and then add food to fill the container. This is recommended because dry ice creates intense cold that might cause plastic to become brittle and crack.

Do *not* use dry ice in glass containers! The pressure might cause the glass to explode.

At one time, Larry and I became energetic and decided to "can" six hundred pounds of wheat all at the same time in our garage. Using five-gallon metal cans, we put in dry ice, poured in the wheat and waited. After a while, we became impatient and decided we had waited long enough. We hurriedly tamped the lids on all the cans. I left, and had gotten part way through the kitchen when I heard a horrible yell. Running back to the garage, I saw Larry standing with his arms thrown over his head as the lids whizzed past him. To this day some of those cans are contoured. We learned that it pays to be patient.

Following is a list of some of the items that can be "canned" at home using dry ice, along with the number of five-gallon cans needed per hundred pounds of food.

Item	No. of 5-gal. cans per 100 pounds
Rolled oats	4
Wheat, rice, whole grains	3
Lentils	3
Beans, peas	3
Soup mix	3
Dried corn	3
Macaroni — elbow or small salad	3
Spaghetti or bulky macaroni	4
Noodles	4
Flour	4
Sugar	4

A fifty-five-gallon drum holds approximately four hundred pounds of wheat.

Dry ice will not affect the sprouting capability of beans or grains. As a seed gets older, its chances of sprouting lessen, but I have sprouted wheat ten years after it was processed with dry ice.

Many commercial companies are using a "nitrogen backfill" process with which to "can" foods. This creates the same type of inert atmosphere as does dry ice.

4

Wheat and Other Cereal Grains

Since ancient times grain has been one of mankind's main foods. Many early peoples believed that the protection of grain was the primary concern of one of the most powerful goddesses, whom the Romans named Ceres. Thus the term *cereal* was derived. Cereals continue to be a major item in the diets of many peoples in many lands.

Cereals include those plants belonging to the grass family which produce grains. This term has now been extended to include food products made from these grains, such as the pastas — noodles, spaghetti and macaroni.

Grains and grain products are good sources of energy at a relatively low cost. Whole grains furnish carbohydrate, protein, iron, and the B vitamins (thiamine, riboflavin and niacin). Protein in cereal is present in varying amounts. A proper combination of grains and the protein of milk, eggs, or meat will meet the body's protein requirements.

Grain Kernels

Kernels of the various grains are individualized, but they are similar to each other in basic structure. Of the many parts to the kernel, only three will be considered here. The wheat kernel is used for illustration purposes.

Bran. The outer covering of the grain is made up of several layers which contain considerable amounts of cellulose, minerals, and protein, and a small amount of thiamine.

Endosperm. The starchy central portion of the grain; also contains some protein.

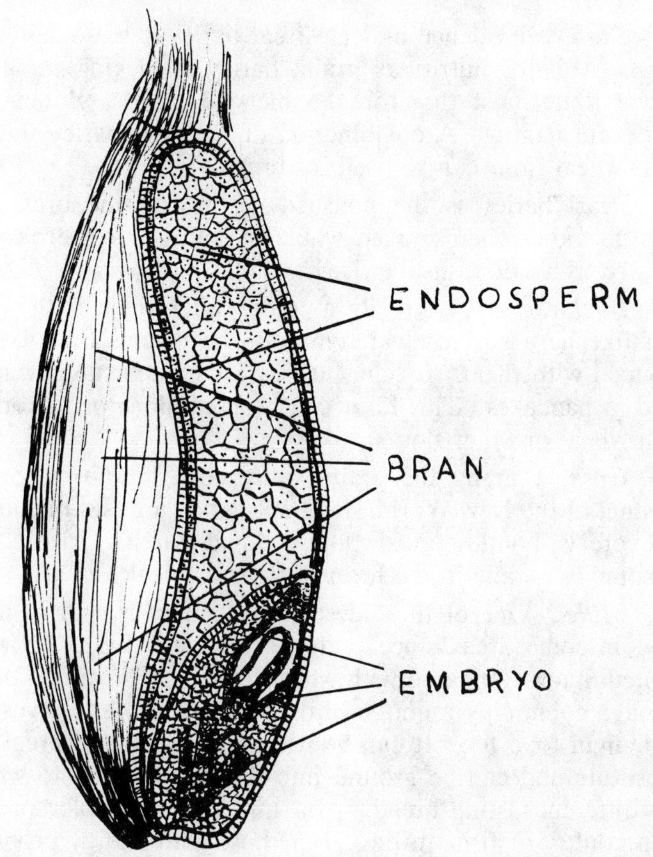

Cross section of a wheat kernel, showing the three main areas where nutrients are found

Germ. Found at one end of the grain, it is actually the embryonic plant. This part of the grain is rich in fat, protein, minerals, thiamine and vitamin E.

The low moisture content of mature grains is responsible for their excellent storage qualities. A moisture content of less than 10 percent is recommended for storage grains.

Grain Varieties and Their Uses

Consider the variety of grains that are available and become familiar with them. You will be surprised how many there are, each with its own special taste and uses.

Barley. Evidence indicates that barley is the oldest cultivated grain. A hard, nutritious grain, barley is lacking in gluten and barley flour must therefore be blended with a gluten flour for successful baking. A combination of one part barley flour to five parts wheat flour can be used for bread.

Pearl barley is the polished grain with the bran removed. It is delicious when cooked whole in soups or casseroles and can be used as a side dish like rice.

Buckwheat. Not really a cereal, buckwheat is the fruit of a grasslike herb so closely resembling the cereals that it is usually grouped with them. Its chief use is for making flour traditionally used in pancakes. This flour can be used alone or in combination with wheat or other flours.

Corn. Among the grains, only corn is known for being a product of the New World. Its great variety of uses includes breakfast cereals, hominy, meal, flour, oil and starch. Dried corn stored at home is usually in the form of flour or meal.

Millet. One of the oldest grains, millet is said to have been used in some areas since prehistoric times. Little known in the United States compared with wheat or oats, it is gaining popularity. Though commonly found in birdseed mixtures, it deserves a chance as human food too. It can be used as a breakfast cereal or a rice substitute and can be ground into flour and added to wheat flour for different-tasting bread. Add millet to casseroles or try using it in poultry or fish stuffing. For those with allergy problems, it is less allergenic than some other grains.

Oats. To many people, the term *oats* means oatmeal, or what is commonly referred to as rolled oats. This can be used as a breakfast cereal or in baking muffins, quick breads, or cookies. Oats are useful as a meat or casserole extender. Oat groats, the whole uncut grain, can be used as a cereal and can be sprouted.

Popcorn. That's right, popcorn! It is a member of the grain family. True, it is not used for breakfast cereal or flour, but it should not be left out of any well-stocked pantry. Winter would be a sad time without a huge bowl of buttered popcorn on a blustery evening. And the Christmas tree would be bleak without popcorn strings. Then the birds might even go hungry, because you can always feed them your popcorn strings after Christmas.

Rice. Rice has been a staple food for many people for centuries. A versatile food, it can be used as a cereal, as a side dish, combined with meat or vegetables in a casserole, or in soups, custards, and puddings. Choose from these five different kinds of rice:

Brown Rice. Whole, unpolished kernels with only the hull removed. It requires more water and a longer cooking time than white rice. Brown rice is not recommended for long-term storage, because the oil in the bran tends to turn rancid over an extended period of time. (One to one and a half years is the maximum storage time recommended. Under ideal conditions it might keep longer.)

Regular White Rice. The hull, bran and germ have been removed by milling. Regular white rice stores well.

Par-boiled or Converted Rice. This has been treated for retention of the natural vitamins and minerals that are contained in the whole grain. It has been cooked before milling.

Pre-Cooked Rice or Quick Rice. Has been completely cooked before packaging.

Enriched Rice. A combination of highly fortified rice and regular milled rice. A coating of vitamins and minerals — thiamine, niacin, iron, and sometimes riboflavin — is used to fortify the rice. Ordinary cooking does not dissolve this coating.

Wild rice is not a true rice, but the seed of a grass that grows in the marshes around the Great Lakes. Very expensive compared with other types of rice, it is said to be of greater food value than cultivated rice. Wild rice is the gourmet item in the rice family.

Rye. As a whole grain, rye is not used extensively in the United States. It is used primarily for making the flour that goes into the familiar rye bread.

Triticale. A relatively new grain, this cross between wheat and rye is used in the same way as wheat — as a breakfast cereal, flour, sprouted, or in many other ways. Triticale flour is excellent for breads and other baked goods. The finished products have a good texture and taste.

Wheat. This is said to be the most important of all the cereal grains. Man apparently grew wheat even before history was recorded. Recorded histories make references to wheat being found all over the world. Today wheat is basic to the diets in many countries, including the United States.

Not only is wheat high in food value and energy value and important for its vitamin content, but it can be stored for long periods of time without deterioration. In addition it is extremely versatile, being used for everything from breads to main course dishes to sweets and desserts. It is valued for flour because of its high gluten content.

Wheat can be used whole or cracked for cereal; in soups, stews, or casseroles; sprouted; and as a meat extender. The possibilities are almost endless. Wheat flour can be used for all baking and cooking purposes. For storage purposes wheat with the best keeping qualities is known as dark hard winter wheat (or turkey wheat). Hard wheat usually has a higher protein content than soft wheat, and contains more gluten.

As with all other grains, wheat should have a moisture content of less than 10 percent if it is to be stored for any length of time.

Basic Cooked Whole Wheat

1 cup whole wheat — 2 cups water
1 tsp. salt

Wash and drain the wheat. Combine wheat and salt in water, and soak overnight (10-12 hours). Place soaked wheat over heat, bring to a boil, boil about 5 minutes, then simmer until tender (1-1½ hours). Makes 8-10 servings. Can be used in casseroles, as a side dish, or any way you wish to serve cooked whole wheat.

Remember that the different grains vary in their content of vitamins, minerals, protein and carbohydrates. If you use a combination of grains in your diet, it provides not only interest for the menu but diversity in the food value of the grains used. This would be especially important in an emergency situation, with a limited variety of food.

Bulgur. An ancient food that has been eaten since biblical times, bulgur is probably man's oldest use of wheat. To make bulgur, wheat is cooked and dried, the outer layers of bran are removed, and the kernels are cracked to the desired size. It looks toasted and has a nutty flavor.

While it is easily stored, dry bulgur should be considered for short-term storage only, because it contains the wheat germ with its oil and will therefore develop a rancid odor if stored in an airtight container. It is properly stored in a dry, cool, well-ventilated location. For storing bulgur a porous container is best (a cardboard box lined with waxed paper would be ideal).

Bulgur is versatile and can be used in many of your favorite dishes. Basically, it may be used in the same way as rice. It can be added to breads, rolls or cookies; served as a cereal; cooked in soups and stews; served as a side dish; or used in whatever gourmet dish your imagination can create. Use it as an extender in meat loaf or chili. For most recipes bulgur must be cooked before use. One caution — it cooks more rapidly than wheat.

Bulgur can be purchased commercially. If you want to make your own, here's how:

Basic Bulgur Recipe[1]

1. Wash wheat in cool water, discard water.
2. Completely cover wheat with water and simmer (about 40 minutes) until water is absorbed and grain is tender.
3. Spread the cooked wheat in a single layer on a cookie sheet and place it in a warm oven (200° F.) to dry. Wheat must be very dry so that it will crack easily. Stir occasionally so that it will dry evenly.
4. When wheat is thoroughly dry, remove chaff by rubbing kernels between the hands. A *little* moisture added either to the hands or to the surface of the wheat will assist in the removal of the chaff.
5. The wheat can be cracked in a mill or grinder (moderately fine). A blender may be used for this purpose.

Storage of Grains

The same basic storage principles apply to all grains. With the exception of bulgur, they should be stored in airtight containers to protect them from insects and rodents. Some grains are purchased in very heavy paper sacks lined with plastic, but this is not sufficient for storage purposes. The grain should be removed from these sacks and placed in airtight containers. (For detailed information on how to "can" grains, and prevent insect infestation of storage foods, see chapter 3.) Be sure that the grains you purchase are clean. This does not mean they have been washed. Washing

[1]Flora Bardwell, *How to Cook and Use Whole Kernel Wheat* (Logan, Utah: Utah State University Extension Service, 1973), p. 9.

would increase the moisture content. Clean grain is free of chaff, dirt, rocks, and as much foreign material as possible. Grains will keep for ten years or longer if properly stored. The storage conditions best suited for grains are outlined in chapter 3.

Uses for Grains

If you aren't familiar with the many grains available, experiment with them. The variety of flavors and textures is great. Try grains in some of these ways:

— As a breakfast cereal, either whole or cracked.
— As a side dish.
— In place of potatoes.
— In combination with vegetables and/or meats, in casseroles.
— For making soups or stews.
— Added to breads for a chewy, textured dough.
— Sprouted and used as a vegetable. (See chapter 6.)

Flour

Flour must be considered in a category by itself, since it is more difficult to store than the whole grain. (For our purposes here we will be referring to whole-wheat flour or white enriched flour.)

Wheat flour is the main ingredient of all breads. The white flour commonly used is ground from wheat, bleached and refined, then enriched with B vitamins and iron. Wheat contains two proteins, gliadin and glutenin, in just the right proportion to make gluten when liquid is added to the flour.

There has been much controversy about the relative food values of enriched white flour and whole-wheat flour. Nutritionists have stated that in milling whole-wheat flour into white flour, the germ and most of the bran are removed. Trace minerals and vitamins are reduced in the white flour. Only thiamine, riboflavin, niacin and iron are added back in enriching the flour. The removed trace minerals are not added back.

The accompanying chart compares the nutritional values of whole-wheat and enriched white flour.

Many people have questioned how rapidly the food value is lost from whole-wheat flour after it has been ground.

NUTRITIVE VALUES OF WHITE FLOUR AND WHOLE-WHEAT FLOUR

Food, Approximate Measure	Pro-tein	Fat	Carbo hy-drate	Cal-cium	Iron	Vit. A	Thia-mine	Ribo-flavin	Nia-cin	Vit. C
	gr.	gr.	gr.	mg.	mg.	I.U.	mg.	mg.	mg.	mg.
100% whole-wheat flour (1 cup — unsifted)	16	2	85	49	4.0	0	0.66	0.14	5.2	0
All-purpose enriched white flour (1 cup — unsifted)	13	1	95	20	3.6*	0	0.55*	0.33*	4.4*	0

*Iron, thiamine, riboflavin, and niacin are based on the minimum levels of enrichment specified in standards set up under the Federal Food, Drug and Cosmetic Act.

Based on information from USDA, *Nutritive Value of Foods,* Home and Garden Bulletin No. 72.

> The vitamin E in whole-wheat flour seems to disappear within hours after grinding. The other vitamins decrease much slower, requiring years. The nutritional value of the protein, carbohydrates, and minerals will last for years. . . . The functions of vitamin E in the human body have not been fully determined.[2]

Whole-wheat flour stored in a *cool, dry* place will keep for approximately one year. The problem in storing whole-wheat flour (at home as well as at the grocery store) is that it contains the germ and oil, which will cause it to turn rancid much more rapidly than white flour. For this reason, wheat flour purchased in the grocery store generally is not 100 percent whole-wheat flour. It is usually enriched white flour with bran added.

White flour will keep for two years or longer if it is stored properly. For best results, use and replace flour so that it will not go stale or become infested with insects. An airtight metal can is the best container for flour storage. It must be kept dark, dry and cool.

The same principles apply to flour made from other grains. If the whole grain is ground into flour, it will contain the germ and oil and cannot be stored for long periods.

---

[2]Osee Hughes and Marion Bennion, *Introductory Foods* (New York: Macmillan Co., 1970), p. 11.

If you reside in an area that has a humid climate it would be well to put flour into airtight containers, using dry ice to prevent insect infestation. (For detailed information on "home canning" flour, refer to chapter 3.)

Gluten

Because gluten comes from flour it is here considered along with the grains. Gluten is an elastic protein substance that gives wheat flour good bread-making qualities. The elastic gluten developed by kneading surrounds and holds the leavening gas produced by fermenting yeast, resulting in light loaves of bread.

Gluten as a food is whole-wheat flour, kneaded well and "washed" to remove the starch. It can be used in making gluten dishes, as a meat substitute, or in a vegetarian diet. Gluten is high in protein and very low in carbohydrates, the carbohydrates having been washed out in the process of "making" gluten.

Work with gluten regularly if you desire to achieve good results. Don't be discouraged with your first few tries. If you plan to rely on gluten as a protein source in an emergency situation, learn to cook with it now.

My first attempts at it were disastrous! Our whole family was willing to at least try a bite of the "hors d'oeuvres," until one fell on the floor and refused to bounce. One loud "blop," then it just lay there. Take my advice and work with it; it gets better.

How to use gluten? There are many ways. It can be sliced and steamed or boiled, then treated as a piece of meat; it can be fried or baked, or ground and used in meat loaf or patties; it can be used as dumplings. Experiment!

Gluten Recipe[3]

7 cups flour (approximately) 2 cups cold water

1. Knead into ball. Knead, beat, pound or stretch the ball of dough for ten minutes. Treat it rough.

2. Wash it. To separate the water-soluble starch from the gluten, cover the ball of dough with cold water and let it stand about an hour. Then run hot water from the faucet over the dough in a strainer. (Catch the water.) Work the dough with your hands

[3]Esther Dickey, *Passport to Survival* (Salt Lake City: Bookcraft, 1969), p. 38.

in the water, washing out the milky starch. As the water becomes milky, pour it off. Keep adding fresh water and working the dough until the water you pour off is clean. (The milky water can be saved for use in bread dough, or soup stock, gravies, etc.)

3. Cook. After the washing process, you will have about 1½ cups of raw gluten. Drain the gluten well and knead into a smooth ball. Try cooking the gluten in a variety of ways.

Gluten may be flavored during the initial mixing by using bouillon, soups, juices, etc., as the liquid. Alternatively, after washing, add spices such as sage, poultry seasoning, or oregano, depending on the kind of recipe the gluten is to be used in.

Pasta

Pasta includes all members of the spaghetti, macaroni and noodle family. Enriched pasta provides useful amounts of thiamine, niacin, riboflavin, and iron, and a small amount of protein. Pasta comes in more than 150 shapes.

Pasta is made from durum wheat, a delicate variety of hard winter wheat. From it comes a high-protein, low-starch flour called semolina, which gives pasta its yellow color and its mild pleasant flavor, and enables it to maintain its texture and shape during cooking.

A 1974 publication brings out some ideas about pasta that may help us put them to better use in our menus.

> Two unfortunate misconceptions still stand between the American consumer and his full enjoyment of pasta. First is the fashionable, if erroneous, belief that flour-based, high-carbohydrate foods are merely "empty calories" and should be eliminated from the diet in favor of high-protein animal products. Responsible nutritionists know that carbohydrates, consumed in moderate amounts, are essential in meeting the body's energy needs. If starches and sugars are absent from the diet, the body uses up protein for energy, thus channeling the latter away from its critical, body-building functions. Furthermore, the U.S. government requires that flour-based products be fortified. They therefore constitute a most important source of B vitamins, iron and calcium. Laws passed in 1973 provide for even greater enrichment of these products, so American pastas are now more nutritious than ever.[4]

[4]"Pasta Past and Present," *Good Foods* (Radnor, Pennsylvania: Triangle Publications, April 1974), p. 12.

NUTRITIVE VALUE OF GRAIN AND GRAIN PRODUCTS

Food, Approximate Measure and Weight 100 Grams — Edible Portion	Calories cal.	Protein gr.	Fat gr.	Carbohydrate gr.	Calcium mg.	Iron mg.	Vit. A I.U.	Thiamine mg.	Riboflavin mg.	Niacin mg.	Vit. C mg.
BARLEY-PEARLED	349	8.2	1.0	78.8	16	2.0	0	0.12	0.05	3.1	0
BUCKWHEAT	335	11.7	2.4	72.9	114	3.1	0	0.60	....	4.4	0
BULGUR Hard Winter Wheat	354	11.2	1.5	75.7	29	3.7	0	0.28	0.14	4.5	0
Corn (Field, whole grain)	348	8.9	3.9	72.2	22	2.1	490*	0.37	0.12	2.2	0
CORN MEAL 1 cup-whole ground (122 gr.)	435	11.0	5.0	90.0	24	2.9	620**	0.46	0.13	2.4	0
GLUTEN-FLOUR	378	41.4	1.9	47.2	40	....	0	....	....	....	0
MACARONI Enriched — dry form	369	12.5	1.2	75.2	27	2.9***	0	0.88***	0.37***	6.0***	0
NOODLES Enriched — dry form	388	12.8	4.6	72.0	31	2.9***	220	0.88***	0.38***	6.0***	0
OATS Oatmeal or rolled oats (dry)	390	14.2	7.4	68,2	53	4.5	0	0.60	0.14	1.0	0
POPCORN (grain)	362	11.9	4.7	72.1	(10)°	(2.5)°	....	(0.39)°	(0.11)°	(2.1)°	0
RICE-BROWN	360	7.5	1.9	77.4	32	1.6	0	0.34	0.05	4.7	0
(MILLED) WHITE	363	6.7	0.4	80.4	24	2.9***	0	0.44***	***	3.5***	0
RYE	334	12.1	1.7	73.4	38°	3.7	0	0.43	0.22	1.6	0
SPAGHETTI (dry enriched)	369	12.5	1.2	75.2	27	2.9***	0	0.88***	0.37***	6.0***	0
WHEAT											
Hard Red Spring	330	14.0	2.2	69.1	36	3.1	0	0.57	0.12	4.3	0
Hard Red Winter	330	12.3	1.8	71.7	46	3.4	0	0.52	0.12	4.3	0
Soft Red Winter	326	10.2	2.0	72.1	42	3.5	0	0.43	0.11	(3.6)°	0

*Based on yellow varieties
**Based on product from yellow varieties
***Based on product with minimum level of enrichment
°Values imputed from similar food

Information calculated from USDA *Composition of Foods,* Agriculture Handbook No. 8.

Problems with Grain Use

If grains, particularly wheat, comprise a major portion of your emergency supply, it is of utmost importance to be aware that whole grains have a laxative effect, due to the fiber of the bran. If our body systems are accustomed only to the "luxury" of refined foods, this could prove to be a very distressing problem. Many people have found a drastic change in the diet to be harsh and irritating, especially for a small child. Having to cope with such a distressing problem would be very depressing during a trying time.

Another problem might be that of allergy. Many people are allergic to large amounts of wheat.

The solution to both of these problems is to begin immediately working whole grains and whole grain products into the family diet, beginning with small amounts, then increasing these amounts as the body learns to tolerate these foods. Learning to use and like the foods you store is an important part of home preparedness.

5

Legumes: Dried Peas, Beans, Soybeans

Beans are among the oldest of foods and today are considered an important staple for millions of people.

They once were considered to be worth their weight in gold — the jeweler's "carat" owes its origin to a pea-like bean on the east coast of Africa.

Beans also once figured very prominently in politics. During the age of the Romans, balloting was done with beans. White beans represented a vote of approval and the dark beans meant a negative vote. Today, beans still play an active role in politics — bean soup is a daily "must" in both the Senate and the House dining rooms in the Nation's Capitol.[1]

Dried beans and peas are excellent home storage items, since they store well and furnish good nutrition. The kinds recommended for storage would be a matter of personal preference. Varieties and usage differ in various parts of the country.

As a group, legumes contain approximately twice as much protein as cereal grains and, *on a per-serving basis,* about half as much protein as lean meat. To get the most nutritional return, the incomplete protein derived from these foods should be combined with the complete protein from meat, eggs or dairy products. Beans and peas are low in fat and high in carbohydrates. They are good sources of iron and thiamine, and contain some riboflavin.

The legume family also includes nuts, which, although a good source of protein, are seldom eaten in large enough quantities to contribute a great deal of protein. Perhaps the old standby, the peanut butter and jelly sandwich, deserves more credit than it is given.

---

[1]USDA, *How to Buy Dry Beans, Peas and Lentils,* Home and Garden Bulletin No. 177 (Washington, D.C.: U.S. Government Printing Office, 1970), p. 9.

Buying Beans and Peas

Most beans, peas and lentils are inspected and graded before or after processing. Grades are generally based on shape, size, color, damage and foreign material. The more uniform the size and color, the higher the grade will be. When buying beans, peas or lentils, consider these points:

Color. The color should be bright and uniform. Dullness or loss of color usually indicates an old product.

Size. The beans, peas or lentils should all be about the same size. Mixed sizes will cook unevenly, since smaller beans cook faster than larger ones.

Defects. The product is inferior in quality if the skins are cracked and broken; if they are full of insect holes; or if the package contains a lot of foreign material such as stems, leaves, dirt and rocks.

Some Varieties Available

You will not find all of the varieties in your area, as favorites vary from state to state. These are some of the more common ones available:

— Black beans. Sometimes called black turtle soup beans. Used in thick soups, Oriental and Mediterranean dishes.

— Black-eye peas. Sometimes called black-eye beans or cowpeas. Small, oval-shaped, cream-colored beans with a small black spot on one side. Used for main dishes.

— Garbanzo beans. Also called chick-peas. Small, nut-shaped, nut-flavored bean used as a main dish vegetable or added to salads.

— Great northern beans. A large white bean used in soups, salads, casseroles and baked beans.

— Kidney beans. A red, kidney-shaped bean used for chili con carne, bean salads, and many Mexican dishes.

— Lima beans. Two varieties — the large lima and the baby lima — both broad, flat, white beans used in casseroles, in soups, baked.

— Mung beans. A small, round green bean used for sprouting. Oriental bean sprouts come from this bean.

— Navy bean. A small white bean. Delicious in soups and baked beans.
— Pea beans. A small, oval white bean used in baked beans, soups and casseroles. Holds its shape even when cooked soft.
— Pinto beans. A beige speckled bean used in salads, chili, and Mexican dishes.
— Red and pink beans. Pink beans tend to have a milder flavor than the red. Used in Mexican food.
— Soy beans. Small, round, green or beige bean. When soaked, it becomes more bean shaped. A very versatile bean.
— Peas. May be green or yellow and can be bought either split or whole. Green has more distinctive flavor; yellow is milder flavored. Split peas have skins removed.
— Lentils. Disc-shaped, flat. Lentils cook in a very short time. Good in soups, stews or casseroles.

If beans become aged (extremely hard) they won't soften by soaking and cooking. If this happens, crack them as you would crack corn or grain. This cracking can be done in a hand grinder or by placing the beans in a heavy paper sack and "cracking" them with the side of a hammer. After cracking, soak and cook them.[2]

Soybeans

Although the soybean is a member of the legume family, it has lately become the aristocrat. It has been a giant in the industrial world for many years, and has been used extensively as animal feed. In today's world of shortages and food problems, the soybean is being recognized as a valuable human food.

Best known for its high protein content, the soybean has about one and one-half times more protein than any other legume. With its quality as well as quantity of protein, the soybean can supplement the protein that we normally obtain from meat sources.

The soybean has many remarkable qualities. From it can be made a milk which resembles cow's milk in many ways. This

---

[2]Barbara G. Salsbury, *Tasty Imitations* (Bountiful, Utah: Horizon Publishers, 1973), p. 71.

milk can be coagulated by acid to form curd — soybean cheese, or tofu. Or if the milk is allowed to ferment, a curd will form.

Numerous recipes are available for the soybean, so no attempt will be made to give complete instructions here. I hope that what is given will whet your appetite sufficiently so that you will not ignore the potential boost to your diet and budget the soybean can bring.

Soybeans can be used as a vegetable, a bean, milk, flour, cheese, pulp, nuts, meat extender or substitute, oil, grits, and sprouts. They have one quirk, however, that other beans do not have. With regular cooking they will never become tender. If you want them soft, they must be pressure cooked.

Cooking Legumes

Cooking dried beans and peas may seem difficult and bothersome to those who have become spoiled with convenience foods. If you are one of those, the aroma and flavor of a pot of homemade bean soup should help overcome apprehensive thoughts about the time involved. Try some and see! Here are the simple basics:

Wash beans, peas or lentils. To reduce the cooking time required, dried beans and whole peas should be soaked before cooking. Split peas and lentils may be cooked *without* soaking.

APPROXIMATE COOKING TIMES — BEANS, PEAS, LENTILS

Kind	Cooking Time
Great northern	1-1½ hours
Kidney	2 hours
Baby lima	45 minutes
Large lima	1 hour
Navy beans	1½-2 hours
Pinto beans	2 hours
Soy beans	1½-2 hours
Green peas	30-45 minutes
Yellow peas	30-45 minutes
Split peas	30-45 minutes
Lentils	30 minutes

Cover beans or whole peas with water, bring to a boil and boil two to three minutes with lid on. Turn off the heat, allow them to soak for one hour, and they are ready to cook.

Remember to allow for expansion of beans, peas and lentils as they cook. They absorb a good deal of water and swell.

Salt and season beans only after cooking, as salt toughens the skin and increases the cooking time.

Approximate Yield

The yield from various beans depends on their size. For example, great northerns will yield more than lentils or split peas. The approximate figures below will serve as a guide.

1 cup dry beans yields 2-3 cups cooked.

1 pound yields 9 servings of 6 ounces each or 12 servings of soup.

Storage Tips

Dried beans and peas store well for a long period of time if moisture content is less than 10 percent and they are kept in airtight containers in a dark, dry, cool place. (See chapter 3 for proper "home canning" methods.) Do not mix newly purchased beans with older "stored" beans. Since older beans take longer to cook, mixing results in uneven cooking — or "bones" in your beans.

6
Sprouting

Sprouting recently has become extremely popular. Many people are familiar with bean sprouts in Oriental food but have not yet had the fun of discovering that sprouting can be done at home. Along with the fun is a very practical side of sprouting.

Why Sprout?

Sprouting enables us to have fresh greens in our diet in any season or weather. It has been said that "growing your own sprouts from seeds is almost like having a vegetable garden in your kitchen. You get a crop rain or shine — easily, neatly, and quickly — in two to seven days."[1]

Sprouting is a nutritious way to use the grains and beans that are stored for a rainy day. Sprouts provide variety in the greens that are needed in our diet and could be especially welcome if we had no other source for these. Any seed, grain, bean or pea can be sprouted. This does not imply that sprouts can or should replace fruits and vegetables, but they are alternate sources of greens that some of us may not be familiar with.

The nutritive value of sprouts has not been officially investigated yet, but private tests have shown that the content of some vitamins increases when a seed is sprouted. Most tests show an increase in vitamin C and a small increase in vitamin A. "Cornell researchers found a fourfold increase in riboflavin and a twofold increase in niacin during the sprouting of soybeans."[2]

---

[1]*Sunset Magazine* (Menlo Park, California), February 1974, p. 64.

[2]Philip S. Chen and Helen D. Chung, *Soybeans for Health and a Longer Life* (New Canaan, Conn.: Keats Publishing Co., Inc., 1973), p. 85.

Sprouts should be considered a supplement to other vitamin sources, as you would have to consume great quantities of sprouts for enough vitamin C to meet the daily allowance recommended by the USDA.

Sprouts — Kinds and Uses

Of two basic types of sprouts, the small seeds and grains such as alfalfa, wheat, caraway, and oats can be sprouted until they form leaves or they can be used sooner according to taste. The larger ones (mung, lentil, alaska pea, pinto, navy) are usually eaten before the leaves open.

Once you become used to the distinctive flavors and characteristics of various sprouts, you can use them as freely as you do celery or onions. Try them as a garnish, as part of a salad, or as the main ingredient in a soup or casserole. They can even be added to waffle batter or bread dough. They can be used raw or cooked, the only exception being soybean sprouts, which should be cooked because they have an anti-digestive factor if used raw.

You might like these other sprout suggestions:

In sandwich fillings
Sprinkled over cheese salads
In meatloaf
Added to pancakes, muffins
In omelets
Raw for nibbling
Steamed and buttered as a vegetable
With fried rice

It has been said that sprouted beans do not produce the gas problems caused by beans. (Note: "It has been said," but not necessarily proven.) Sprouted beans will cook faster and have an excellent flavor. Cooking time is reduced to thirty or forty minutes. The whole sprout, including the seed, is edible. Some people prefer to remove the hulls from some types of seeds, but it is not necessary.

Equipment

You don't need any fancy or special equipment. Once you get started you will discover you can sprout in virtually anything.

Large or small, it's still fun! As you see, most of the following suggestions include common household items.

- — A shallow casserole dish, pie plate, or serving dish covered with cheesecloth will work. Well-used old bath towels or dish towels serve well for sprouting.
- — Cookie sheets or sheet cake pans can be used.
- — One person has suggested using 46-ounce juice cans or 3-pound shortening cans for sprouting the larger beans. Perforate the bottom of the can with small holes. After soaking, the beans can be rinsed and left to drain and sprout in this container.
- — A large colander or strainer makes an excellent sprouter.
- — A quart-size jar is very handy. For good drainage, use a canning jar ring to hold a piece of nylon stocking, cheesecloth, old terry cloth or screening over the opening.
- — A screen frame can be made easily by fastening fiberglass screening over a wooden frame. If several frames are made to stack conveniently, various stages of sprouting can be kept in one compact area.
- — Several commercial sprouters are available. Many health food stores carry them.

Methods

Sort through the seeds to be used for sprouting and discard any broken, cracked or unhealthy looking seeds. If this is not done, the broken seeds will decay and mold.

Cover the seeds with warm (not hot) water and soak them for ten to twelve hours. Allow enough room and water for expansion, remembering that seeds and beans increase three to four times their original volume. Small seeds become saturated much sooner than the larger, tougher ones.

Drain off the excess water and keep the seeds in a warm place. If a jar is used for sprouting, the seeds can be soaked in the sprouter jar. For better growth lay the jar on its side. After rinsing, drain the water through a screen or cloth covering.

When sprouting in dishes or trays, keep the sprouts covered with a clean damp cloth to hold in the moisture and warmth. Keep the seeds moist by rinsing or spraying with lukewarm water three to four times a day. If the sprouts are rinsed under the faucet, take care that the force of the water does not break the sprout from the seed. Be sure to drain after rinsing.

Good drainage, a *must* with sprouting, is as important as proper moisture. After rinsing, allow the sprouts to sit in the sink or on the drainboard for a short while to prevent any problems from water drainage.

After the sprouts have started to grow, light will cause them to turn green. If a white sprout is desired, keep them in the dark. For the small green leafy kind, by all means put them in the light.

Mung bean sprouts: Actual size

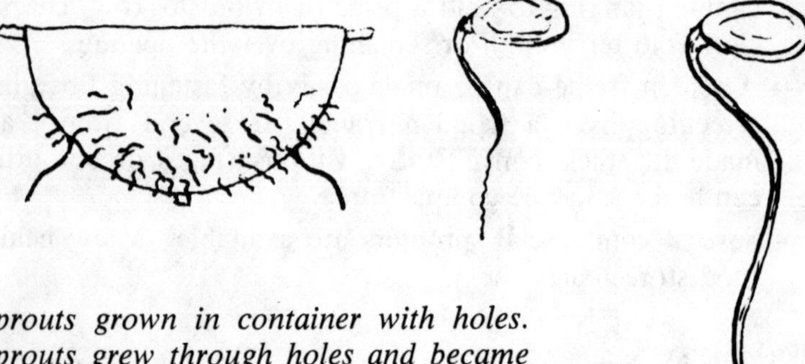

Sprouts grown in container with holes. Sprouts grew through holes and became shriveled and brown. Did not grow properly.

Sprouts in jar grew to desired length. All of sprout was good to eat.

One problem is noted in using a sprouting container that has holes in it, such as a colander or perforated sprouter. Sprouts from the larger beans such as mung beans grow through the holes. Very soon a good growth is retarded and the end of the sprout shrivels. It appears that the circulation gets cut off and the sprout cannot grow properly. If this happens, it takes only one day's growth to ruin the whole batch. A gallon jar works much better for batches of big beans.

Sprouts will keep well for several days if refrigerated in a covered container or a plastic bag.

Even younger children can be involved in sprouting projects. Being "put in charge" of the sprouts can help them gain a sense of success and accomplishment. You will find, too, that if they grow it, they will be more willing to eat it.

Experiment now so that you become familiar with how to sprout and what to do with your "crops." As a starter, try sprouting some of these:

Alfalfa	Brown rice
Mustard	Wheat
Radish	Fenugreek
Caraway	Lentil
Garden cress	Soybean
Mung bean	Pinto
Alaska pea	

7

Home Preservation of Foods

Preserving foods at home is exciting, satisfying, rewarding and the results taste oh, so good! Not that it isn't work, but when we compare our up-to-date methods and equipment with the struggles and efforts of our grandparents, we have it pretty easy. To fill those shelves is a major task. Peeling gets tiring. Standing over a hot stove in the middle of summer takes more than courage. Cleaning up the kitchen afterward is a wearisome task. And then a miracle takes place! For when those shelves are full of beautiful, gleaming jars of fruits and vegetables we have a great feeling of joy and accomplishment. With the freezer full we have fewer worries. Best of all is the satisfaction that comes from having on hand year-around a supply of good, nourishing food. I have often sat on a small stool in my fruit room with tears of true thanksgiving streaming down my face.

Is It Practical?

Before methods and applications are explored, let's discuss whether home preservation is practical, especially for those of us who don't live in a farm area.

If you have even a small plot of ground, grow your own produce. Yes, even in the big cities. Look for "how-to" suggestions in chapter 12.

Where to Buy

Following are a few suggestions on where and how to obtain produce for preserving at home.

In many large cities, during the summer season, you will see produce trucks that have good buys. By ordering the day before and making advance preparations, you can process the fruit or vegetable as fresh as possible. Prices of produce on these trucks are often lower than in the supermarket.

The produce managers in your local supermarkets can be a big help to you. Ask them when the fruits or vegetables that interest you will be in best supply. Let them know that you will be interested in buying in quantity. Have a friend or two join you on the purchase, in order to get reduced rates for buying several lugs.

Let these same produce managers know of your interest in "spotted or bruised" produce. If used promptly there is nothing wrong with a bruised peach, but it will be thrown out by the market. It doesn't take many bruised peaches (minus the bruises) to fill several quart jars, or to make a batch of good-tasting preserves.

Another produce source for city dwellers is the farmers' market, where produce dealers buy from the growers. There is usually an area open to the public, and prices are good. Plan to buy by the lug or larger amounts, not by the pound. Farmers' markets open very early in the morning, usually by 4:00 A.M. If you like hustle and bustle and enjoy the aroma of fresh-picked fruits and vegetables, this is a rewarding experience!

Small produce stands, usually in the outlying areas, are another source for fresh fruits and vegetables.

In areas where fruits and vegetables are grown there are growers who will allow you to come in and pick your own. The saving is usually worth the effort, but the experience of gleaning the fields even surpasses the saving. Though you get grubby and tired and your muscles ache, it is worth it. The warm earth smells good and the fruit tastes good. Tastes? Oh yes, you always eat some as you pick. That's part of the fun.

Season for Preserving

Summer is not the only time to preserve fruits and vegetables. Many are in season at other times, depending on where you live. Some of the crops harvested in the fall will keep for a while: squash, pumpkins, beets, apples, pears, etc. Preserve immediately the perishable ones, such as tomatoes and peaches; the others can be done a little at a time. The time to preserve food is when you can get good quality produce at reasonable cost.

Other Costs

What about costs other than for produce? Do they negate the thrift aspects of home preserving?

If you have the time, and produce can be obtained at reasonable prices, there is no question about the economy, let alone the good quality of home-preserved foods. But your circumstances and time schedule will determine the extent of your home preserving. Perhaps your time does not permit a large-scale operation, but you could easily handle small quantities.

Sugar or Honey

January, February and March are the best months in which to buy at least the bulk of your sugar or honey for canning and freezing, since sugar usually goes on "special" then. During the remainder of the year buy one or two bags each shopping day. This will insure ample supply and a spread of cash outlay. Sugar prices are usually highest during canning season. If you prefer to use an artificial sweetener such as saccharine, the same principles apply. It is best to plan for July and August in December and January.

Bottles, Lids, Rings

Don't wait until canning season to purchase bottles. If you have canned in previous years you will have some "empties," but for initial supply or renewals try the local thrift store instead of the supermarket. Why pay "new" prices for jars you have to wash anyway? A used dirty jar washes just as easily as a new one, but costs considerably less. When buying used jars, be sure to check the rims for chips or cracks. Used jars are likely to be more abundant in mid-winter. Check garage sales for good buys on bottles. Buy any new jars, lids and rings during the winter, when the supply is usually plentiful. Harvest-time demand often brings a short supply.

Throughout the year, save mayonnaise, pickle, and jelly jars for relishes, sauces, jellies and preserves. Use only regular canning jars for the hot water bath or pressure cooker. Ordinary jars will break very easily under heat or pressure, but they are fine for open-kettle sauces and jams.

FRESH FRUIT AND VEGETABLE AVAILABILITY

Produce Item	January	February	March	April	May	June	July	August	September	October	November	December
Apples	X	X	X	X	X	O	O	O	X	X	X	X
Apricots	†	†	†	†	O	‡	‡	†	†	†	†	†
Bananas	X	X	X	X	X	X	X	X	X	X	X	X
Beans, Snap	O	O	O	O	X	X	X	X	O	O	O	O
Beets	O	O	O	O	O	X	X	X	X	X	O	O
Broccoli	X	X	X	X	O	O	O	O	O	X	X	X
Cabbage	X	X	X	X	X	X	X	X	X	X	X	X
Cantaloup	†	†	†	†	X	‡	‡	‡	X	X	†	†
Carrots	X	X	X	X	X	X	X	X	X	X	X	X
Cauliflower	O	O	O	O	O	O	O	O	X	X	X	O
Celery	X	X	X	X	X	X	X	X	X	X	X	X
Cherries	†	†	†	†	O	‡	‡	O	†	†	†	†
Corn, Sweet	†	†	O	O	X	X	X	X	X	O	O	†
Cucumbers	X	X	X	X	X	X	X	X	X	X	X	X
Grapefruit	X	X	X	X	X	O	O	O	O	O	X	X
Grapes	†	†	†	†	O	O	X	‡	‡	‡	X	O
Greens	X	X	X	X	X	X	X	X	X	X	X	X
Lettuce	X	X	X	X	X	X	X	X	X	X	X	X
Onions	O	O	X	X	X	X	X	X	X	O	O	O
Oranges	X	X	X	X	X	X	O	O	O	X	X	X
Peaches	†	†	†	†	O	X	‡	‡	X	†	†	†
Pears	O	O	O	O	O	†	O	X	X	X	X	O
Peppers	X	X	X	X	X	X	X	X	X	X	X	X
Pineapples	O	O	X	X	X	X	O	O	O	O	O	O
Plums - Prunes	†	†	†	†	†	O	‡	‡	X	O	†	†
Potatoes	X	X	X	X	X	X	X	X	X	X	X	X
Squash	X	X	X	X	X	X	X	X	X	X	X	X
Strawberries	O	O	X	‡	‡	‡	X	O	O	O	O	O
Sweet Potatoes	X	X	X	X	O	O	O	O	X	X	‡	X
Tomatoes	X	X	X	X	X	X	X	X	X	X	X	X

Key to chart:

- † Supply scarce or nonexistent
- O Moderate supplies available
- X Supplies plentiful
- ‡ Peak of season — supplies abundant

Freezing Supplies

For similar price and supply reasons, buy plastic freezer bags and boxes off season.

A little preplanning along with wise buying of produce for canning and freezing will help to cut the budget while preserving foods in a variety of ways at home.

The chart on page 47, adapted from one by the United Fresh Fruit and Vegetable Association, shows when fruits and vegetables are at their seasonal peak and the most abundant. Prices should be the best at those times. The chart shows averages only, since seasons vary in different parts of the country. Abundance of crops can vary with weather patterns.

8

Home Canning

Whether you are a veteran home canner or a novice, your approach is important. Depending on your approach, you may find canning an enjoyable challenge or an unbearable chore; an exciting and rewarding project or tiresome drudgery. If you have never canned before, go easy and your enthusiasm will carry you through the whole season. It is much more enjoyable to process a few jars at a time rather than several lugs. In your eagerness to do all of these new things, take care not to bite off more than you can chew. (This advice applies to all of us.) A few jars at a time add up quickly, while overzealous buying can result in waste, spoilage and frustration.

Start plans for the canning season several weeks ahead of time. Decide the kinds and amounts of canned foods you need and want, including your family's favorite fruits and vegetables. Take stock of canned items already in the storeroom.

Most of the needed canning equipment is already in your kitchen.

CANNING EQUIPMENT

Jar lifter
Tongs
Wide-mouth funnels (one or two)
Large bowls or dishpans
Large ladle
Sharp knives
Colander
Scrubbing brush
Chopping board
Peeler
Food strainer
Wire basket
Trays
Large measuring cups (several)
Thick hot pads (several)
Towels and dishcloth
Large saucepans
Forks and spoons

Helpful additions:

Blender | Food chopper | Household scale

Acceptable Home Canning Methods

The boiling water bath is used for *acid* foods such as fruits and tomatoes. Even pickles in brine are insured a longer life if processed a few minutes in the water bath. Jars packed with raw food are placed in a kettle deep enough to allow water to cover the jars. The water is then boiled for a specified time, according to jar size and type of food. Enough heat is supplied by the boiling water to destroy bacteria, enzymes, molds and yeasts, which cause spoilage in acid foods. A water bath canner may be purchased at hardware stores, catalog order stores and department stores, and generally come in either the seven-quart or the nine-quart size.

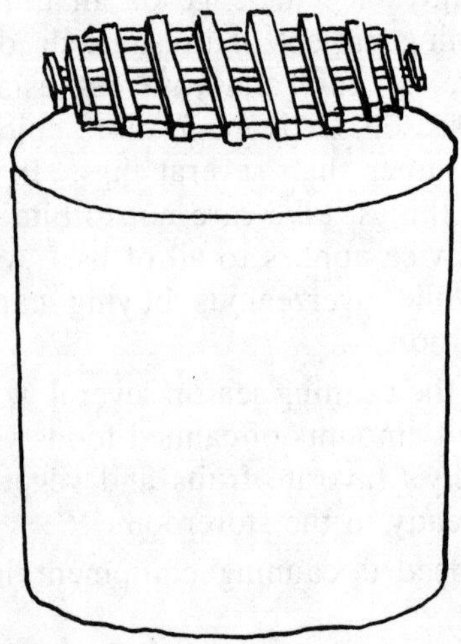

Homemade waterbath canner

Any very large pail or kettle may be used, if it is deep enough so that the water will come at least one inch over the top of the jars. Allow additional room at the top for the water to boil. The kettle must have a wooden or sturdy wire rack that will hold the jars off the bottom of the canner. The rack must be open enough to allow the water to circulate, but solid enough so that the bottles will not fall against one another and break as the water boils.

Be sure to follow the times shown on the processing tables exactly.

With a pressure cooker, foods are processed under pressure at a temperature of 240° F. This canning method, for all nonacid foods such as meats and vegetables, gives a greater degree of safety from food spoilage, especially the types that cause food poisoning.

A pressure canner is a heavy kettle with a lid which can be clamped or locked on to make it steam-tight and which has a safety valve and a pressure gauge. All parts of this type of canner *must* be in good working order. The pressure gauge should be checked annually and adjusted for altitude. Be sure to follow carefully the manufacturer's instructions for operating the kind of pressure cooker being used.

Containers — Cans or Jars

Tin cans. These are widely used in homes where a sealer is owned. While the sealer is an expensive piece of equipment initially, it pays to have one if canning is done on a large scale. The sealer automatically seals the disc-like tin lids onto the open top of the tin can. The same can may be used as many as three times, since the sealer can reflange the open top and seal it again, thus extending the usefulness of the can and reducing the expense of the cans.

It is more economical to purchase cans in large quantities. They are generally available at local hardware stores. Two of the most popular sizes are: no. 2, which holds approximately 2½ cups, and is generally preferred for fruits and vegetables; no. 2½, which holds approximately 3½ cups, and is generally used for canning meats.

With tin cans, advantages to be considered include elimination of breakage, cans are sealed before processing, heat penetration is more rapid than with glass, and the cans can be cooled quickly by plunging them into cold water.

Glass jars. The sizes generally used for home canning are the quart, pint and half-pint, all of which can be obtained with either wide-mouth or regular openings. The size of jars used would be determined by the individual or family need.

New lids are required for self-sealing jars each time the jar is reused. The self-sealing lid and ring is the most popular type in use today.

Canning Check List

These general rules of procedure apply, no matter which method you use — open kettle, water bath or pressure cooker:

1. Use only good quality, fresh, clean fruits and vegetables. Dirt may harbor organisms that cause spoilage.

2. Sort jars, rings and lids. Check jars for chips and cracks. Never use a jar that has "only a small crack." Your fruit or vegetable as well as the jar will be lost. Slivers of glass could get into the food from cracked or chipped jars. If the rim of a jar is cracked or chipped, *it will not seal*.

The rubber (sealing edge) around the outside of the lids must be solid and show no deterioration. Rings (old or new) should be clean and free from food or corrosion. *Before* you start the canning process, make sure you have sufficient supplies on hand — enough rings and lids for the jars; and enough jars for the amount of fruit.

3. *Read* the recipe first! Have all of the necessary ingredients on hand.

4. When canning fruit, make the syrup in advance and keep it hot.

5. Wash all produce carefully. To prevent loss of juice, stem berries and cherries after washing. Prepare vegetables as if for cooking, and peel fruits and tomatoes (except for cherries and small apricots, which can be canned whole).

6. When filling jars, work as quickly as possible, packing one jar at a time.

Canning Process

Pack food in jars so that it neither crowds nor wastes space. You can pack firmly and neatly without mashing the product. A jar that is packed too full may not seal properly. Packing too tightly may also prevent proper heating. Except when using the open kettle method, *always leave headroom at the top of the jar*.

Fruits and leafy vegetables do not expand very much and so require only one-half inch of headroom. Vegetables such as string beans, lima beans, peas and corn expand much more and need one inch of headroom.

The liquid should barely cover the food; otherwise, with the expansion that takes placing during processing, liquid may be forced out over the rim and the lid will not seal.

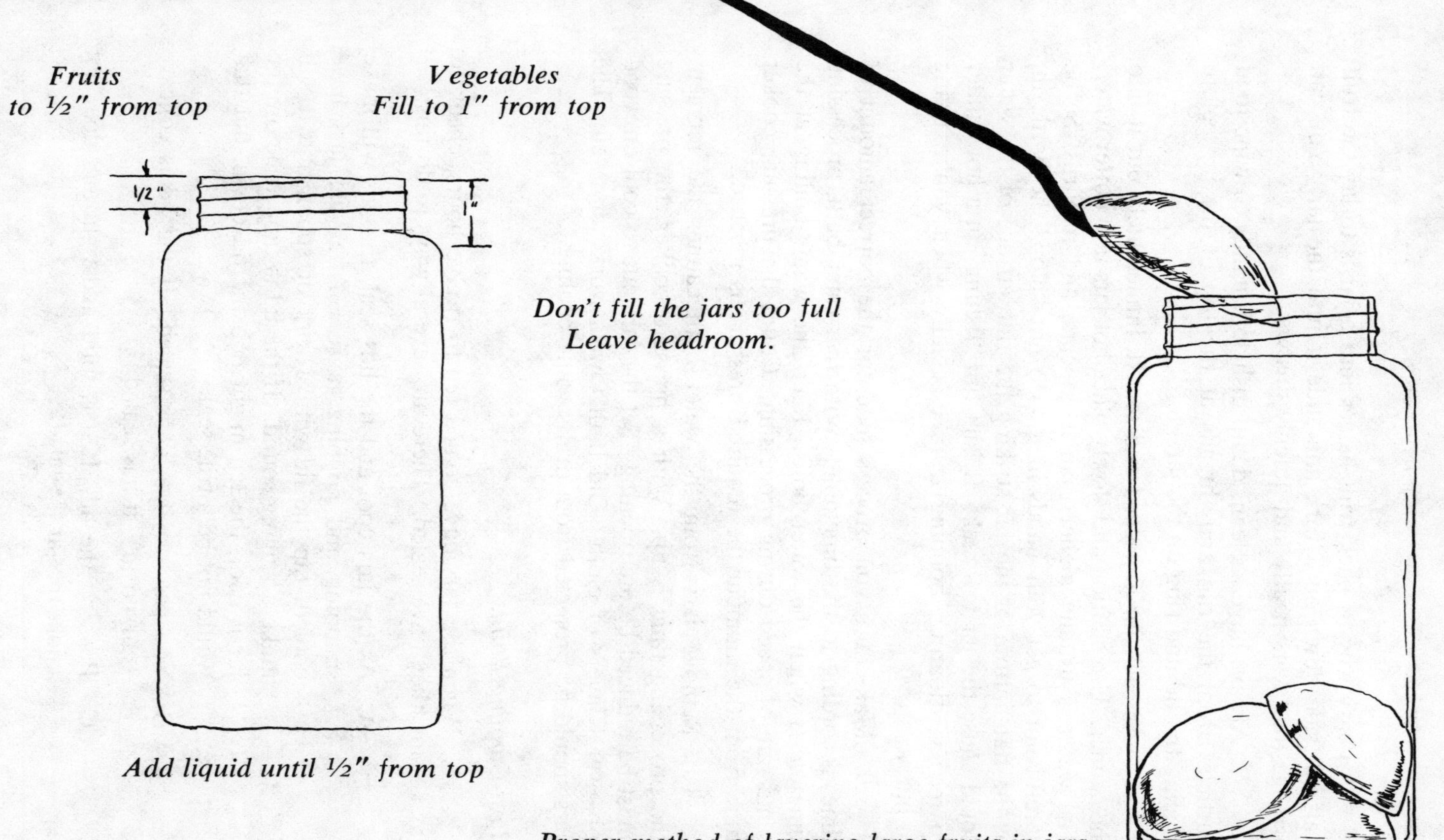

Proper method of layering large fruits in jars

An easy way to pack fruit halves into a jar is to lay the fruit half cut side down on a fork and slide it into the jar. You can arrange the layers neatly and quickly this way.

7. Always have a clean, damp dishcloth handy to wipe food particles from the rim and threads at the top of the jar before placing the lid and ring on the jar.

8. As each jar is packed and sealed, immediately put it into the canner. Use caution and slowly place hot jars into hotter water. The water is usually higher than it appears, and it burns! So be careful not to get your hands in it. Do not "plop" the bottle in, as the flat bottom of the jar striking the flat surface of the water could cause the jar to break. It isn't fun "fishing" in a hot canner to retrieve floating, bobbing fruit, especially when you're trying to hurry.

9. Make sure you process food for the correct amount of time, according to the timetable. With the water bath, processing time *starts* when the water boils. Keep the water boiling at the same rate of speed during processing. Count the pressure cooker time from the moment the required pressure is reached.

10. Remove jars from the water bath canner the moment the processing time is up. With a pressure cooker, the pressure must be reduced before opening. Set the jars out to cool on wood or several layers of cloth. Old bath towels work very well. The jars should be protected from drafts while cooling.

Post-Canning Points

1. When the jars have cooled ten to twelve hours, check to make sure they have sealed. There are several ways to do this:

A. As the jars cool and the lids seal, the lids will make snapping and popping noises. After the jars have cooled, tap the lid with a fork or spoon and it will make a ringing sound. If the lid is tapped before the jar is cool, the tap might pop the lid down, but that would not be a true seal.

B. You can see if a lid is sealed. If the lid is concave (curved in), it has sealed.

C. Push on the lid. If it is solid and tight and does not click, the jar is sealed.

2. After cooling jars for twenty-four hours, wipe them clean with a damp cloth, then label and date them. Jars that show any signs of not sealing should be opened and the contents heated, repacked, and reprocessed.

3. Store jars in a dark, dry, cool place. Light causes foods to fade and destroys vitamins.

4. If you want to check whether a jar has remained sealed, before opening it, pour one teaspoon of water on the lid; then puncture the lid with an ice pick or can opener. If the water is sucked down — "slurp" — the jar has remained sealed.

5. *Never taste to test for spoilage!* If a food looks spoiled, has an unnatural foaming, or an off odor during heating, destroy it.

6. Before tasting or serving any home canned meat, fish, poultry or nonacid vegetables, boil them for ten to fifteen minutes uncovered. If the tomatoes you can are low-acid, this instruction also applies to them.

Open Kettle Method

The open kettle method is recommended only for food products that contain a large amount of vinegar or sugar, since these products help to preserve them. Fruit and tomatoes can be processed in this way if other methods are not available, but this method is more suited for jams, preserves, sauces, pickled foods, relishes, and some pickles. Follow these steps:

— Use hot sterilized jars. Wash and rinse the jars in hot water, then turn them upside down on the oven shelf, with the oven set on its lowest temperature. A standard oven keeps twelve jars ready. Or wash jars in the dishwasher, then maintain it at a warm setting.

— Place lids in a pan of *hot* water, ready for use. Boiling is unnecessary.

— Make a light syrup, add fruit, and cook until tender. If it is to remain fruit, don't cook it too much.)

— Work right next to your stove so that the fruit can be kept boiling as jars are filled.

— Using a wide-mouthed funnel, fill a single jar at a time. Work quickly.

— Remove the funnel, and insert a table knife along the sides of the jar to disperse air bubbles.
— Wipe the rim and top of the jar, then put on a hot lid and screw the ring on tightly.

Caution: Oven canning is considered very dangerous. Not only is this method unreliable because foods may spoil, but jars can explode when put in the oven.

Methods for Packing Jars

With the *raw or cold pack* method, you pack unheated or raw food into the jar, then add boiling liquid — syrup, water or juice — leaving 1½ inches head space. The jars are sealed, then processed. Most fruits and tomatoes are best if packed raw.

The *hot pack* method utilizes a short precooking. Boiling hot food is then packed into clean jars, sealed and processed immediately. The hot pack is more satisfactory for some vegetables and meats that will be processed in the pressure cooker. Large fruits such as pears and peaches can be either cold or hot packed. Fruits such as apples and rhubarb are best if hot packed. Fill jar with syrup, juice or water, leaving a half-inch head space.

Canning Without Sugar

All fruit and fruit juices may be canned without sugar, but "canning in sugar helps to improve flavors of most fruits. The sugar does not keep the fruit from spoiling."[1] "When canning fruit without sugar, the jar is filled within a half-inch of the top with water or fruit juice and the hot pack method of canning is recommended."[2]

Canning for Sugar-Free Diets

Hot Pack: Preheat fruits over low heat in a small amount of water. Pack hot fruits and cover with juice from cooking kettle. Process according to timetable.

---

[1]*Kerr Home Canning Book* (Sand Springs, Oklahoma: Kerr Glass Manufacturing Co., 1972), p. 13.

[2]Betty M. West, *Diabetic Menus, Meals and Recipes* (Garden Citv. New New York: Doubleday & Co., 1949), p. 201.

Raw Pack: Pack fruit raw (cold). Add fruit juice to within 1½ inches of top of jar. To obtain fruit juice, crush thoroughly ripe fruit and bring to a boil over a low heat. Strain through a clean cloth. Process jars with water or fruit juice same time as fruits packed with syrup.[3]

Sweeteners Other Than Sugar

If you plan to use artificial sweeteners in canning, read and follow the manufacturers' instructions for proper use.

Where honey or corn syrup is used in place of sugar, it is recommended that no more than half of the sugar be replaced with light honey or light corn syrup for fruit canning. (Dark syrup or honey has a stronger flavor.) If more is used, the syrup or honey flavor may overpower the flavor of the fruit.

In jams, jellies and preserves, up to a third of the sugar can be replaced with corn syrup. Up to one-half of the sugar can be replaced with honey.

Sugar Syrups

Use the syrup that best fits your family taste.

Thin: 1 cup sugar to 3 cups water.

Medium: 1 cup sugar to 2 cups water.

Heavy: 1 cup sugar to 1 cup water.

The sweeter and riper the fruit, the less sugar you need. If you find the fruit too tart for your taste when you open the jar, sugar can be added then.

Some people prefer to put the sugar directly into the jar (one to two teaspoons per quart), adding hot water to fill it about one-fourth full, stirring to mix well, then putting in the fruit. Additional hot water can be added if needed. You may find this a faster method.

Don't Forget the Salt

Two teaspoons of salt should be added to each quart of canned vegetables. If you put salt in the jar, you can see it. If you leave it until last, you may forget it or salt twice.

---

[3]*Kerr Home Canning Book,* p. 13.

Discoloration

In order to keep peeled fruit from turning brown during preparation, drop it in a mild solution of 1½ teaspoons salt to a quart of cold water.

Ascorbic acid is a powdered substance that will help prevent discoloration. Add ¼ teaspoon to each quart of fruit before filling the jar with syrup. Although discoloration does not harm the food value or change the taste of the fruit, it does affect the appearance.

WATER BATH PROCESSING TIME IN MINUTES

Fruit	Kind of Pack	Half-Pints	Pints	Quarts
Apples	Hot	15	20	20
Apricots	Raw	20	25	30
Apricots	Hot	15	20	25
Berries	Raw	10	15	20
Berries	Hot	10	10	15
Cherries	Raw	15	20	25
Cherries	Hot	10	10	15
Mixed fruit	Hot	15	20	25
Nectarines	Raw	20	25	30
Nectarines	Hot	15	20	25
Peaches	Raw	20	25	30
Peaches	Hot	15	20	25
Pears	Hot	15	20	25
Pineapple	Hot	10	15	20
Plums	Hot	15	20	25
Rhubarb	Hot	10	10	10
Tomatoes	Raw	30	35	45
Tomatoes	Hot	10	10	15

Courtesy of *Ball Blue Book,* El Monte, California.

PRESSURE COOKER PROCESSING TIME IN MINUTES

Vegetable	Kind of Pack	(Steam Pres. Canner [240° F.] 10 Pounds of Pressure) Half-Pints and Pints	Quarts
Asparagus	Raw or Hot	25	30
Beans (green)	Raw or Hot	20	25
Beans (lima)	Raw or Hot	40	50
Beets	Hot	30	35
Broccoli	Hot	30	35
Brussel sprouts	Hot	30	35
Cabbage	Hot	30	35
Carrots	Raw or Hot	25	30
Cauliflower	Hot	30	35
Celery	Hot	30	35
Corn, whole kernel	Raw or Hot	55	85
Corn, cream style	Hot	85	Not Recommended
Greens, all kinds	Hot	70	90
Peas, green	Raw or Hot	40	40
Potatoes, sweet	Hot and Wet	55	90
Pumpkin	Hot	65	80
Turnips	Hot	30	35

Note: The time given in these tables applies to sea level and must be increased for altitudes of 1,000 feet or more. For each 1,000 feet above sea level, add 1 minute to processing time if the time called for is 20 minutes or less. If the processing time called for is more than 20 minutes, add 2 minutes for each 1,000 feet.

When a pressure cooker is used at an altitude of 2,000 feet or more, the pressure must be increased by 1 pound for each 2,000 feet of altitude.

Courtesy of *Ball Blue Book,* El Monte, California.

Estimating Yields

The exact number of jars needed for a particular food depends on the size and condition of the produce and the way it is prepared and packed into the jars. The standard weight of a bushel basket, lug or box may not be the same for all fruits, but the accompanying chart can be used as a guide to determine the approximate yield per bushel or lug of produce.

APPROXIMATE YIELD OF CANNED FRUITS AND VEGETABLES FROM FRESH PRODUCE

Raw Produce	Fresh	Canned
Apples (sliced)	1 bu. - 48 lb.	16-20 quarts
	2½-3 lb.	1 quart
(sauce)	1 bu. - 48 lb.	15-18 quarts
	2½-3½ lb.	1 quart
Apricots	1 lug - 22 lb.	7-11 quarts
	2-2½ lb.	1 quart
Berries	24 quart crate	12-18 quarts
	1½-3 lb.	1 quart
Cherries	1 bu. - 56 lb.	22-32 quarts (unpitted)
	2-2½ lb.	1 quart
Peaches	1 bu. - 48 lb.	18-24 quarts
	1 lug - 22 lb.	8-12 quarts
	2-3 lb.	1 quart
Pears	1 bu. - 50 lb.	20-25 quarts
	1 box - 35 lb.	14-17 quarts
	2-3 lb.	1 quart
Plums	1 bu. - 56 lb.	24-30 quarts
	1 lug - 24 lb.	12 quarts
	1½-2½ lb.	1 quart
Tomatoes	1 bu. - 53 lb.	15-20 quarts
	1 lug - 30 lb.	10 quarts
	2½-3½ lb.	1 quart
Beans, lima	1 bu. - 32 lb.	6-10 quarts
	3-5 lb.	1 quart
Beans, green	1 bu. - 30 lb.	12-20 quarts
	1½-2½ lb.	1 quart
Beets (no tops)	1 bu. - 52 lb.	15-24 quarts
	2-3½ lb.	1 quart
Carrots	1 bu. - 50 lb.	16-25 quarts
	2-3 lb.	1 quart
Corn, sweet (in husks)	1 bu. - 35 lb.	6-10 quarts
	3-6 lb.	1 quart
Peas, green (in pods)	1 bu. - 30 lb.	5-10 quarts
	3-6 lb.	1 quart
Spinach (other greens)	1 bu. - 18 lb.	3-8 quarts
	2-6 lb.	1 quart
Squash (summer)	1 bu. - 40 lb.	10-20 quarts
	2-4 lb.	1 quart
Sweet Potatoes	1 bu. - 50 lb.	16-25 quarts
	2-3 lb.	1 quart

Why Canned Foods Spoil

Principal reasons for spoilage of canned foods are:

— Poor quality food and delay in canning
— Jars not properly washed and cleaned
— Too little processing time
— Imperfect seal (rims cracked or chipped or not properly cleaned)
— Careless storage

Enemies of Successful Food Preservation

Preserving food at home isn't difficult, but no one wants to spend time, money and effort doing so if the foods will spoil and be wasted. If we understand what causes canned foods to spoil and follow a few rules, we can avoid spoilage. Basically, it involves doing two things: 1) stopping enzyme action; 2) killing small organisms such as bacteria, yeast and mold.

Enzymes are present in all fresh foods. They are responsible for the normal ripening of fruits and vegetables. Once the ripening process is complete, the enzyme action must stop or it will cause decay. Extreme heat or extreme cold will stop enzyme action.

Bacteria, yeast and mold are present at all times in soil, water and air. Extreme heat or cold will kill these organisms, but the food must then be kept in a sealed, airtight container to prevent further contamination.

Processing, the application of heat to food after it is packed in jars, is done to stop the enzyme action and to kill bacteria, yeast and mold that are present in foods.

More Ideas

Slipping skins on peaches, apricots or tomatoes saves time. Put the washed fruit into a wire basket or cloth bag and drop it into enough boiling water to completely cover it. After almost one minute, remove fruit from the hot water and plunge it into cold water to stop the cooking process. The skins will slip off easily if the fruit is ripe.

In order to get your money's worth, don't discard fruit pits and peelings. They can be boiled and the juice strained and made into jelly. Don't throw anything away. Fruit that has spots or bruises can be trimmed and sliced. Small pieces, even spotted or bruised pieces, can be used for preserves or butters. Naturally you would not use wormy or diseased fruit.

Extra work space can be obtained by setting up a card table. By using this space to cool the jars, regular table and counter space are free for other work.

It's wise to teach young children how to prepare foods, but canning season is a time to be careful. Let children pick, sort, wash, peel and chop foods. Avoid accidents by keeping them away from all processing equipment.

Emergency Processing

In an emergency situation you could can by a water bath or open kettle method on a camp stove, wood or coal stove, or even over an open fire. You would need to take greater safety precautions and the process would take longer. And you would have to keep the fire as constant as possible.

Storage

Foods canned by the open kettle method do not have as long a shelf life as those that are cold packed, so they should be dated and used first. While some foods hold their color, flavor and texture longer than others, changes generally can be noticed within a year. With jellies, jams, preserves and butters, the shorter the storage time the better the flavor.

Now you should be ready to can!

9
Home Freezing

If the pioneers and pilgrim fathers could have foreseen the modern marvels we use with such ease, they surely would have envied us. Just think of eating strawberries in the middle of January! Or savor the thought of sweet, juicy peaches or melon in the middle of a blizzard! Sound delicious? There is no "out of season" with a freezer.

There is an even more practical side of having "round the year" choice of fruits and vegetables. Freezing is a safe, easy way to preserve almost any food at home, including fruits, vegetables, meat, poultry, fish, baked goods and casseroles. Frozen foods, generally prepared for the table before freezing, can be counted on as a budget booster when winter prices rise, as well as being convenient to use. Freezing as a method of home preservation has the advantages of keeping the natural color, flavor and food value of most foods.

Choose Carefully

Be selective in your choice of produce and meats to be frozen. If properly handled and packaged, frozen food can be preserved in the same good condition, but freezing does not improve the quality. Choose solid, ripe produce.

What to Freeze

What foods to freeze must be decided on an individual or family basis. Needs, likes and dislikes, freezer space, budget, and other storage space all are determining factors.

For example, I prefer to can most of our fruits, sauces, relishes and jams, and freeze the vegetables and meats. Since moving to

Utah my family have enjoyed experimenting with and using a root cellar along with fruit and vegetable dryers. It all depends on personal preference and circumstances.

Some varieties of all fruits and vegetables freeze better than others. Growing conditions vary a great deal in different parts of the country. Contact your local state extension service or experiment station or college of agriculture for information on which varieties would be best in your area.

If you have doubts about how well a fruit or vegetable will freeze, or how you will like the taste of it after it has been frozen, try a small amount first. Freeze an amount that will be eaten at one meal, then use these foods right away to see if you are pleased with the results. Do this before you invest time, money and effort in a larger batch. This would be especially wise with foods that are new to you. No matter how low the price, it's not a bargain if you throw it out later because you don't care for it. Not only would the food be wasted, so would the freezer space that is filled with unwanted items.

In an Emergency

Many people are concerned about the possibility of a power failure, and perhaps the loss of a freezer full of food. There is always the chance of a blown fuse, a broken fan or a disconnected cord. The energy crisis threatens us with possible power shortages, but at this point it looks as though electricity is here to stay. Nevertheless, perhaps we need to be aware of precautions to take if the power is interrupted or the freezer fails to operate for a short period of time. There are several ways to protect the "investment" in the freezer.

— Keep the freezer closed. If full, the contents should remain frozen two days; if half full, one day.

— For a longer period, use dry ice to prevent thawing. Fifty pounds of dry ice will last three to four days in a twenty-cubic-foot freezer.

— If the freezer is partially loaded, move all of the packages to a central location, placing the dry ice on the center of them.

— Use caution in handling dry ice — it burns. Wear gloves or use heavy paper.

What and When to Refreeze

If some thawing takes place, all is not lost. Some foods may be refrozen. If foods have thawed only partially and there are still ice crystals in the packages and they are firm in the center, they can be refrozen.

All uncooked foods can be cooked and frozen again. Some items, such as fruits, can be made into preserves and canned.

> Because low acid foods, which include most of the vegetables, spoil rapidly after they have thawed and warmed into temperatures above 45°F., it is generally not advisable to attempt to refreeze them.[1]

Fruit and fruit products can be refrozen if they are still very cold. Their flavor and quality may be lessened but they will be safe to eat.

Wrapping Materials and Containers

Improper packaging is the main cause of deterioration in frozen foods. Since freezer air is very cold and dry, badly wrapped foods lose moisture, flavor and quality. Ordinary household foil, clear wrap or waxed paper, cottage cheese cartons and ice cream cartons should not be used, as they cannot protect food properly. Food to be frozen should be put in packages that will keep it from drying out and prevent air from causing the foods to turn dark.

> All containers used must be airtight, leakproof, moisture-proof, odorproof and vaporproof if you want to retain the highest quality of the food to be frozen. It should be strong and easy to fill, close and empty.[2]

The type of packaging to be used for a particular product depends on the size and shape of the food, its consistency, and whether or not it is a solid or liquid. It is economical to choose containers which can be reused.

Rigid containers. Rigid containers made from glass, aluminum or plastic are needed for liquid packs of fruit and for combination dishes which contain liquid. These containers must have tight-fitting lids.

---

[1]USDA, *Home Freezing of Fruits and Vegetables* (Washington, D.C.: U.S. Government Printing Office, 1954).

[2]USDA, *Home Freezing,* p. 4.

Regular glass canning jars may be used for freezing, but as frozen food containers, they have some disadvantages. It takes several hours to thaw the food enough to remove it from the jar, because the jar-size piece of frozen food cannot pass through the mouth of the jar. Canning jars take up a great deal of space also, and they are difficult to stack.

Nonrigid containers.

> Bags and sheets of moisture-vapor-resistant cellophane, heavy aluminum foil, pliofilm, polyethylene, or laminated papers and duplex bags consisting of various combinations of paper, metal foil, glassine, cellophane and rubber latex are suitable for dry packed vegetables and fruits.[3]

Bags come in a variety of sizes and are easy to use. They are most satisfactory with dry packed vegetables, and raw poultry or meat. While bags can be used for liquid packs, they are not as convenient as a rigid container. When closing a freezer bag, press the air out, twist the end, and fasten with a rubber band, wire twist or freezer tape.

The various freezer papers or materials that come in sheets or on rolls are good to use for fish, poultry, meat, or corn on the cob. Freezer-weight aluminum foil is very good for wrapping irregularly shaped items such as a chicken because it can be molded close to the food to keep out the air, yet it is strong and remains flexible at low temperatures.

Regular cellophane becomes very brittle and breaks easily at freezer temperatures.

In order to obtain the maximum use from any container you decide to use, follow the manufacturer's suggestions.

Cost and the budget. Compare the prices of the available containers. Compare the convenience of the various types, and the space needed to store them, and consider whether or not they are reuseable. Perhaps a higher initial cost may mean a greater saving in the long run.

Packaging

All food should be cold when it is packed. To eliminate as much air as possible, pack the food tightly. Allow ample head-

---

[3]USDA, *Home Freezing*, p. 4.

space because the food expands as it freezes. Keeping the sealing edges of containers clean will help to obtain an airtight seal.

Make it a habit to accurately label containers immediately before they are put into the freezer. After food is frozen, it is hard to identify it. Date each package.

Headroom. Remember that food pushes up as it freezes, and the upward pressure could force off a lid or break a bag. If a canning jar is used, the pressure could cause the glass to break unless proper headroom is allowed.

Do not use canning jars as freezer containers for fruits or vegetables that are packed in water.

Vegetables that can be packed loosely, such as asparagus or broccoli, do not require any headroom.

HEADROOM FOR FREEZING FRUITS AND VEGETABLES

	Wide-Top Container		Narrow-Top Container	
Type of Pack	Pint	Quart	Pint	Quart
Liquid Pack: Fruit packed in juice, syrup or water; crushed or pureed; soups or stews	1½ in.	1 in.	¾ in.	1½ in.
Juices: fruit juices, vegetable juices, broth or stock	½ in.	1 in.	½ in.	1½ in.
Dry Pack: Fruits or vegetables packed without sugar or liquid	½ in.	½ in.	½ in.	½ in.

Some foods can be frozen before packaging. Such foods as berries, peas, lima beans, sliced or diced green pepper, and chopped onion are more convenient to use if frozen loose, and it is very simple to do. To use this method on green peppers, simply slice them onto a cookie sheet in a single layer and place the cookie sheet in the freezer until the peppers are frozen. Then put frozen pepper slices into a bag or other container, seal, and place the package in the freezer.

Foods that are frozen this way will not stick together and come out in a lump, but can be poured out in the desired amount when you are ready to use them.

Fruit

General Preparation

All fruit needs to be thoroughly cleaned. Because ripe fruit bruises easily, wash only a small batch at a time. Using a colander or wire basket, place the fruit under gently running water or plunge it up and down in a sink full of cool water. Drain the fruit promptly. Do not allow it to stand for any length of time in the water.

Small fruits and berries can be frozen whole; larger fruits are usually halved or sliced. Even slightly bruised fruit can be sliced and frozen. Most fruits and berries can also be crushed, pureed or made into juices for freezing.

Packing the Fruit

Fruit is usually packed one of three ways, depending on the type of fruit and how it will be used. Fruits to be used for desserts are usually packed in syrup, while the dry or unsweetened pack is better for cooking.

SYRUPS FOR FREEZING FRUIT

Type of Fruit:	Percent	Sugar (cups)	Water (cups)	Yield (cups)	
Mild flavored — sweeter fruits	30	2	4	5	(light)
	35	2½	4	5⅓	`
Most fruits	40	3	4	5½	(medium)
	50	4¾	4	6½	
Very sour fruits	60	7	4	7¾	(very
	65	8¾	4	8⅔	heavy)

As a general rule, up to one-fourth of the sugar can be replaced with corn syrup. A larger proportion of corn syrup can be used if it is light colored and bland in flavor.

Hull, pit or peel the fruit, working with only enough to fill a few containers at a time. Prepare fruit the way you would for serving, keeping it as cold as possible. If it is scalded or steamed, plunge it into ice water to cool it before packing.

If fruit is to be crushed, a blender or potato masher can be used. For pureed fruits, a blender, colander or strainer works well.

Preventing Browning

Some fruits such as apples, apricots, peaches and pears turn brown when the fruit is peeled or cut. To prevent this darkening process, use ascorbic acid powder (vitamin C), generally about one-fourth teaspoon to two cups of syrup. Ascorbic acid comes in both crystalline and tablet form, and usually can be purchased wherever canning and freezing supplies are sold and at most drugstores. Follow label directions when using ascorbic acid.

While it is not absolutely necessary to use ascorbic acid when freezing fruits, it does help them to retain their natural color. It also adds a little to the expense of freezing.

Dissolve ascorbic acid in a little cold water. If tablets are used, crush them so that they will dissolve more easily.

For use in fruit that is syrup packed, dissolve the acid in cold water and add it to the cold syrup just before packing the fruit. When freezing fruits that darken quickly, mix ascorbic acid with the syrup and slice the fruit directly into the syrup.

In dry sugar packed fruit, dissolve the acid in cold water and sprinkle it over the fruit before adding the sugar. If fruit is being packed without sugar, dissolve the acid in cold water and sprinkle it over the fruit before packing.

Make syrup by dissolving the sugar in cold or hot water. If hot water is used, be sure to cool the syrup before packing fruit in it. Syrup can be made the day before and refrigerated.

For the *syrup pack,* put the fruit in the container and cover it with cold syrup. Be sure that syrup covers all the fruit. Any uncovered top pieces will change color and flavor. A piece of waxed paper can be crumpled in the top of the container to hold the fruit under the syrup.

With the *sugar pack,* sugar is sprinkled over the prepared fruit. Mix it thoroughly but gently. A large wooden spoon is better than a fork for this purpose and your hands can do the job even more gently. Put fruit into the containers and seal.

For the *unsweetened pack,* simply pack the fruit into the containers.

Fruit Freezing Hints

Allow ample headroom. Pack and label according to intended use. Included here are suggestions for individual fruits.

— Apples. Use firm, crisp fruit. Use syrup pack for fruit to be used as uncooked dessert. For baking or cooking, use unsweetened or dry sugar pack. Apples darken easily.
— Apricots. Scald for one minute, then cool in ice water, to prevent the skins from becoming tough. To serve as uncooked fruit, use a syrup pack. For cooking purposes, use a sugar pack.
— Berries. To serve uncooked, pack in a medium syrup, or freeze with a sugar pack or unsweetened.
— Cherries. To serve as uncooked fruit, use a syrup pack; for cooking purposes, a sugar pack.
— Cranberries. Pack unsweetened with a heavy syrup pack.
— Fruit Cocktail. Pack in a medium syrup.
— Melons. Pack balls or cubes in medium syrup.
— Peaches. Use a medium syrup or a dry sugar pack. Peaches brown easily, so use ascorbic acid.
— Pears. Pack with medium syrup. Since pears darken quickly, use ascorbic acid.
— Pineapple. Pack unsweetened or use light syrup.
— Plums. Use either a dry sugar pack or a medium syrup.
— Strawberries. Use either a sugar or syrup pack. Small berries are better frozen whole; large berries are better sliced or crushed.

How to Use Frozen Fruit

To serve raw, thaw in the unopened container, then serve as soon as thawed. Fruits served when they are still icy have a better texture. Don't thaw more than you plan to serve at the time. Any leftover thawed fruit will keep better if it is cooked. Fruit will keep several days in the refrigerator after it is cooked.

To cook with frozen fruit, thaw it until the pieces come apart easily, then cook with it as you would fresh fruit. If a recipe calls for sugar, remember to allow for any that was used in packing the fruit. Frozen fruit may have more juice than a recipe calls for. Be sure to measure the fruit before adding it.

Crushed fruits and purees can be used for toppings, baby food, fillings, puddings, sherbets, pies, special diet foods, and in fruit-filled cookies and cakes.

YIELD OF FROZEN FRUITS FROM FRESH

Fruit	Fresh	Frozen
Apples	1 bu. - 48 lb.	32-40 pints
	1 box - 44 lb.	29-35 pints
	1¼-1½ bu.	1 pint
Apricots	1 bu. - 48 lb.	60-72 pints
	1 lug - 22 lb.	28-33 pints
	⅔-⅘ lb.	1 pint
Berries	1 crate 24 quarts	32-36 pints
	1⅓-1½ pints	1 pint
Cantaloup	1 dozen - 28 lb.	22 pints
	1-1¼ lb.	1 pint
Cherries	1 bu. - 56 lb.	36-44 pints
	1¼-1½ lb.	1 pint
Cranberries	1 box - 25 lb.	50 pints
	1 peck - 8 lb.	16 pints
	½ lb.	1 pint
Currants	2 quarts - 3 lb.	4 pints
	¾ lb.	1 pint
Peaches	1 bu. - 48 lb.	32-48 pints
	1 lug - 20 lb.	13-20 pints
	1-1½ lb.	1 pint
Pears	1 bu. - 50 lb.	40-50 pints
	1 western box - 46 lb.	37-46 pints
	1-1¼ lb.	1 pint
Pineapple	5 lb.	4 pints
Plums or Prunes	1 bu. - 56 lb.	38-56 pints
	1 lug - 20 lb.	13-20 pints
	1-1½ lb.	1 pint
Raspberries	1 crate - 24 pints	24 pints
	1 pint	1 pint
Rhubarb	15 lb.	15-22 pints
	⅔ to 1 lb.	1 pint
Strawberries	1 crate - 24 quarts	38 pints
	⅔ quart	1 pint

From USDA, *Home Freezing of Fruits and Vegetables.*

Frozen Jam

Using commercial pectin, some ripe fruits make delicious frozen jam. Here's how:

Frozen Uncooked Jam

Mix 3 cups of crushed blackberries, blueberries, raspberries, strawberries or peaches with 5 cups of sugar; let stand 20 minutes, stirring occasionally. Dissolve 1 package of powdered pectin in 1 cup of water, heat to boiling and boil for 1 minute. Pour pectin solution into fruit mixture and stir for 2 minutes. Put into jelly glasses or freezer containers. Seal.

Since uncooked jams spoil very easily, they must be kept frozen or refrigerated.

VEGETABLES

Only vegetables which are usually cooked before serving should be considered for freezing. Salad vegetables such as lettuce, celery, radishes, cucumbers, cabbage, and green onions do not stay crisp or appetizing after freezing.

Vegetables should be thoroughly washed in cold water, cut up and sorted according to size.

Blanching

All vegetables except green peppers should be blanched (scalded) before freezing to stop the action of the enzymes. (Before vegetables are picked, enzymes help them to mature; after picking, enzymes cause loss of flavor and color.) Work quickly, with small quantities at a time. Have a large container and plenty of ice cold water ready for chilling the vegetables. If you are blanching a large quantity of vegetables, change this water frequently in order to keep it cold. For blanching, use about a gallon of boiling water per pound of vegetables. Prepare the vegetables as you would for cooking or canning. Put two to three pints of the prepared vegetable in a wire basket or cheesecloth bag. The time required for heating a vegetable in boiling water varies with the size of the vegetable. The water should be heated to a good rolling boil before you put the vegetables into it. Start timing as soon as you put vegetables in the water.

SCALDING TIME IN MINUTES

Vegetable	Minutes
Asparagus stalks	2-4
Beans, lima (or pods)	2-4
Beans, green or wax or french cut	3
Beets	25-50 (until tender)
Broccoli stalks (split)	3
Brussel sprouts	3-5
Carrots (whole)	5
Carrots (diced or sliced)	2
Cauliflower	3
Corn on the cob	7-11
Corn — whole kernel or cream style, cut corn from the cob after cooling	4
Peas — green	1½-2
Spinach	2
Squash, summer	3

Note: If you live 5,000 feet or more above sea level, blanch one minute longer than the time specified above.

From USDA, *Yearbook of Agriculture,* 1959.

Steaming

Some vegetables can be heated by steaming. Split broccoli and whole mushrooms take 5 minutes; button mushrooms 3-5 minutes; and sliced mushrooms 3 minutes. Pieces of pumpkin and winter squash and whole sweet potatoes can be steamed until soft or they can be heated in a pressure cooker or an oven set at 400° F. until they are soft.[4]

For steaming, use a kettle with a tight lid or a rack that holds a steaming basket at least three inches above the bottom of the kettle. Put in one to two inches of water and bring to a boil. Place the vegetables in the basket in a single layer. The steam needs to reach all parts of the vegetables as quickly as possible. Cover the kettle and maintain a high heat. Start timing when you put the lid on. If you live above 5,000 feet, steam one minute longer than the recommended time.

Cooling Is Important

Quick, thorough cooling is necessary. Take the vegetables from the blanching kettle and plunge the basket or cloth bag directly

[4]USDA, *Yearbook of Agriculture* (Washington, D.C.: U.S. Government Printing Office, 1959), p. 465.

into a large amount of cold water — below 60° F. The water used must be cold enough to stop the cooking process. Ice water or cold running water is good to use. It will take approximately as long to cool the vegetables properly as it did to heat them.

Packaging

After draining them well, pack the cooled vegetables immediately into suitable containers. Leave one-half inch headroom, except for vegetables like asparagus and broccoli that pack loosely and do not require any headroom.

APPROXIMATE YIELD OF FROZEN VEGETABLES FROM FRESH VEGETABLES

Vegetable	Fresh, as Purchased or Picked	Frozen
Asparagus	1 crate — 12 2-lb. bunches 1-1½ lb.	15-22 pints 1 pint
Beans — lima (in pods)	1 bu. - 32 lb. 2-2½ lb.	12-16 pints 1 pint
Beans — snap green or wax	1 bu. - 30 lb. ⅔-1 lb.	30-45 pints 1 pint
Beet — greens	15 lb. 1-1½ lb.	10-15 pints 1 pint
Beets — without tops	1 bu. - 52 lb. 1¼-1½ lb.	35-42 pints 1 pint
Broccoli	1 crate - 25 lb. 1 lb.	24 pints 1 pint
Brussel sprouts	4 qt. boxes 1 lb.	6 pints 1 pint
Carrots	1 bu. - 50 lb. 1¼-1½ lb.	32-40 pints 1 pint
Cauliflower	2 medium heads 1⅓ lb.	3 pints 1 pint
Corn	1 bu. - 35 lb. 2-2½ lb.	14-17 pints 1 pint
Peas	1 bu. - 30 lb. 2-2½ lb.	12-15 pints 1 pint
Peppers	⅔ lb. - 3 peppers	1 pint
Pumpkin	3 lb.	2 pints
Spinach	1 bu. - 18 lb. 1-1½ lb.	12-18 pints 1 pint
Squash	1 bu. - 40 lb. 1-1¼ lb.	32-40 pints 1 pint
Sweet Potatoes	⅔ lb.	1 pint

USDA, *Home Freezing of Fruits and Vegetables.*

Vegetable Freezing Hints

— Asparagus. Break off tough parts of the stem (save them for making soup). Cut or break into pieces two to four inches long.

— Beans. Green or wax, leave whole or snap into half-inch pieces.

— Lima beans. Sort according to size.

— Beets. If small, can be frozen whole; larger ones can be sliced or diced. The greens freeze easily.

— Broccoli or cauliflower. If heads are too large, split them. If necessary to remove insects, soak for half an hour in a salt solution (four teaspoons salt to a gallon water), wash and drain.

— Corn on the cob. Prepare the same as for serving.

— Corn, whole kernel. Blanch corn on the cob four minutes, cool and drain, then cut kernels from the cob. Don't cut too close to the cob.

— Corn, cream style. Cut the corn from the cob at about the center of the kernel, then scrape the cob with the back of the knife to remove the juice and the heart of the kernel.

— Greens. Prepare as for serving.

— Peas. Shell and discard hard, overripe peas. Peas are easy to freeze in a single layer on a cookie sheet.

— Pumpkin or winter squash. Cook and mash, as for pies.

— Sweet potatoes. Cook until almost done, but not overdone or they will go mushy.

— Turnips. Use young tender turnips. Older ones have a much stronger flavor.

How to Use Frozen Vegetables

For best flavor, texture and nutrition, cook frozen vegetables in a small amount of water only until they are tender. Long cooking is unnecessary as they have already been blanched. Overcooked frozen vegetables will have a mushy texture.

Put frozen vegetables into boiling water and cover the pan. Remember to add seasonings. Cooking time will vary with the vegetable.

If you are serving an oven-cooked meal, frozen vegetables can be baked in a covered casserole. Just place the frozen vegetable in the greased casserole, add butter or margarine and season to taste. Cover and bake until tender.

Use frozen vegetables as you would the fresh ones.

Meat, Poultry, Fish

Freezing is a very good way to preserve meat, fish or poultry. For city dwellers it is a good way to help overcome the high cost of meat. With a freezer you can take advantage of seasonal or promotional meat sales by making quantity purchases when supplies are available and the price is right. For the successful fisherman, that catch of fish can be kept rather than given away.

However, before you buy half a beef, do a little comparing. Know that the price is for the entire half and find out how many pounds of waste bone and fat will be thrown out when it is cut and trimmed. About 25 percent is lost in fat and bone. Consider whether you can save money by buying only the cuts of meat that you prefer when they are on sale. There will probably be an additional cost for cutting and wrapping, unless you plan to do it yourself. Even considering these costs, there may be a good savings on buying meat in quantity compared to purchasing weekly at the local market. One advantage of buying half a beef is the convenience — it's there when you need it.

General Preparations

Meat to be frozen should be of high quality. Keep clean all food that is to be frozen. It should be prepared and ready to cook before freezing. Package in quantities that you normally cook at one time.

Very large quantities of meat, such as a half a beef or venison, are usually cleaned, chilled, aged, cut and wrapped by a professional butcher.

Prepackaged Meats

When you purchase cuts of meat prewrapped in the clear plastic wrap at the local supermarket, they should be cared for properly. If the meat will be used within a few days and will be kept in the freezer compartment of the refrigerator, it need not be rewrapped. However, if a large quantity of meat is purchased for the freezer, it does need further care. The store wrap is not durable enough to protect the meat at freezer temperatures. Remove the store wrap and rewrap tightly in proper freezing materials. It will be worth the extra time and effort.

Poultry can be frozen whole, in halves, or in pieces, depending on how it will be used. Unless you know it will be used whole, it is much easier and more convenient to cut up poultry prior to freezing. Poultry intended for broiling, frying or barbequing will take less freezer space and be more convenient to use if it is cut into serving pieces before freezing.

Caution: Do not attempt to stuff poultry before freezing. Dangerous bacteria can develop in poultry stuffed at home and then frozen, and normal roasting or baking will not kill these bacteria. Large commercial companies sell pre-stuffed poultry, but they work under strictly controlled conditions not possible in home kitchens.

Poultry should be wrapped tightly to protect it from drying out, and frozen immediately. It is usually more convenient to wrap and freeze the giblets separately.

Fish should be washed and cleaned and waste portions removed before packaging. Packaged this way, fish take less storage space than the whole fish. As with other meats, package fish in the amount needed to serve one meal.

Wrapping and Packaging

All meat, poultry and fish should be frozen in moisture-vapor-resistant materials to make the package airtight and prevent drying. *Freezer burn* is a term applied to foods that have been improperly packaged so that exposure has caused them to dry to the point where color, flavor and texture are destroyed.

In preparing meat for freezing, place two layers of waxed paper between individual chops, steaks and fillets. This makes for convenience and faster thawing and makes it easier to separate individual frozen pieces. Ground meat may be packaged in cartons or bags, or shaped into loaves or patties before wrapping. Place two layers of waxed paper between individual patties.

Whole chickens are conveniently frozen in freezer bags. For convenience in using cut-up chicken, place waxed paper between the pieces, then use freezer bags or cartons, or wrap tightly and seal.

Packaging of fish is usually determined by the size and kind of fish. Trout, for instance, is usually frozen whole, several to a package; sea bass or other larger fish are generally filleted or made into steaks. Some large fish may be frozen whole if they are to be baked. Whatever the package size, fish must be wrapped very tightly in a durable freezer material.

Wrapping steps.

1. Put the food in the center of the sheet and bring opposite sides of the sheet together.

2. Pull paper tight to drive out air. This also keeps the packages smooth so that they fit together better.

3. Fold the edges down in a series of folds, making a tight seam, and press wrapping tightly against the food.

4. Fold the ends in tightly. Tape ends and seams securely with freezer tape.

Labeling. Label every package before it goes into the freezer, giving the kind and cut of meat, number of servings, the date it went into the freezer, and the date by which it should be used.

Thawing

> It's best to thaw frozen fish, poultry or meat in the refrigerator. That way the surface does not reach dangerously high bacteria levels before the product thaws out in the center. Another benefit of slower thawing is less moisture loss (drip). Meat, fish and poultry can be cooked without thawing; allow about one-third to one-half more cooking time.
>
> Some foods, especially meats and poultry — though purchased thawed — may have been previously frozen. If refrozen, what happens to their quality? Not much as long as they were commercially frozen.[5]

Baked Goods

Most baked goods freeze easily. Prepare ahead or take advantage of bakery thrift stores that sell day-old goods.

— Cakes. All kinds can be frozen; they are better unfrosted. Thaw cakes three to four hours at room temperature; cupcakes fifteen to twenty minutes.

— Cookies. Many kinds can be frozen as dough or after baking. The crisp kind are usually frozen as dough. Macaroons and meringue types don't freeze well.

[5]*Spotlight on Freezer Storage* (Ohio State University Extension Service, 1973).

— Pies or pastries freeze well, baked or unbaked.
— Bread and rolls store well frozen.
— Bread or yeast doughs can be frozen, but should be used within six weeks.

Miscellaneous Ideas

— Freeze leftover bread slices, then cut up for croutons or toppings.
— Frozen shredded cheese can be used for cooking.
— Freeze leftover pancakes or waffles. Reheat in the toaster or in the oven five minutes at 350° F.
— Cook double batches of your favorite dishes, eat some, freeze the rest.
— Undercook dishes that will need to be cooked again, such as casseroles.
— Leftover gravy can be frozen and used later as a soup base.
— Hard-cooked egg whites become rubbery when frozen.
— To freeze leftover whipped cream, drop by spoonsful on waxed paper, freeze, then place pieces in freezer bag. To use, place on fruit or dessert, and allow to thaw a few minutes before serving.

10

Fruit and Vegetable Drying

History

Drying, the oldest method of food preservation, goes back thousands of years. In the early days of man's existence, he was controlled by his food supply. Life was a continual struggle, for there was no known way of preparing for the future. Since man did not understand the causes of food spoilage, he was restricted to those preservation methods that were discovered accidentally.

Early man discovered that food left exposed to the sun and wind dried and could be stored and used in the future. Centuries before the Christian era, appreciation of dried fruits had developed so that dates and grapes were dried in the sun and stored away for winter use.

Corn and fish were dried by early American colonists. The American Indian and early settlers in the arid midwest of the United States dried buffalo meat and beef by cutting these meats into strips and hanging them in the sun to dry. Ancient man dried food in caves and found that heat from the fire hastened the process. During the Civil War, troops were supplied with dried vegetables, apples and peaches. Dried foods were also used as rations during World War I and the Boer War. During World War II the production of dehydrated foods became increasingly important.

Modern Dried Foods

Today nearly all of the fruits and vegetables found in the supermarket can be preserved by drying. Drying foods has remained a popular preservation method with many people because it is a comparatively simple process and requires little outlay in equipment, time and money.

> Successful drying depends upon the removal of enough moisture to prevent spoilage. This must be done as quickly as possible at a temperature that does not seriously affect the texture, color, and flavor of the vegetable or fruit. If the temperature of the air is too high and the humidity of the air too low, there is a danger that moisture will be removed from the surface of the food more rapidly than water can diffuse from the interior and a case hardening will form on the food. This layer will then not permit free diffusion of moisture from inside and the product will not dry properly.[1]

Nutritive Value of Dried Foods

The protein, carbohydrates, and minerals in dried foods remain practically unchanged, but the effect of drying on vitamins is more serious.

> The vitamin content of fruits is changed in drying, depending upon the methods of drying and sulfuring. Some fruits are subjected to the fumes of sulfur dioxide gas to prevent darkening of color and to act as an insecticide. Sulfuring aids in the preservation of vitamins A and C, but it affects thiamine (vitamin B) adversely.[2]

Research indicates that fruits dehydrated indoors with controlled heat and circulation retain more of their vitamins than when sun drying is used.

Dried, Dehydrated, Sun Dried

The three terms, *dried, dehydrated* and *sun dried* are very often used interchangeably, but there are differences between the three.

Dried. Those foods containing only 10 to 20 percent water, the balance of the water having been removed in a piece of equipment designed for home use, using artificial heat.

Dehydrated. Those foods containing only 2.5 to 4 percent water, the other 96+ percent having been removed using equipment too sophisticated to have in the home.

---

[1]Flora H. Bardwell and O. K. Salunkhe, *Home Drying of Fruits and Vegetables* (Logan, Utah: Utah State University Extension Service).

[2]Hughes and Bennion, *Introductory Foods.*

Sun-dried. Foods containing only 10 to 20 percent water. These foods are usually placed on trays and left in the sun until evaporation of the water takes place.

Only two of the methods will be considered in this chapter — sun drying and drying.

Selection

Fruits and vegetables selected for drying should be sound, fresh, and at the right stage of maturity. Drying neither improves nor lowers food quality. Wilted or inferior produce will not make a satisfactory dried product. Immature fruits and vegetables are weak in color and flavor. Overmature vegetables are usually tough and woody.

> The riper the fruit is, the more sugar it will contain and therefore the larger the yield of dry fruit will be, unless the fruit is over-ripe and so soft that excessive loss occurs.[3]

Handling

Any food to be dried, except meat, must be washed carefully. Sort and remove any spoiled or defective produce. Decay or mold on one slice of fruit or one piece of vegetable may cause a trayful of bad flavor. After washing, the food may be peeled if necessary.

Drying must be rapid enough that spoilage will not take place during the process. If the fruit is large, such as peaches, cut it into slices one-half inch thick.

Preparation of Vegetables

Vegetables should be steamed or blanched before drying. This stops the enzyme action, checks ripening, and prevents change after drying.

Steaming. Use a large kettle and a wire basket or sieve so that steam circulates freely around the vegetables, but they do not touch the water. The water should be boiling briskly before putting in the prepared vegetables.

Layer the vegetables no more than 2½ inches deep. Steam until each piece is thoroughly heated and wilted. Test by removing a piece from the center of the steamer and pressing it. It should feel tender but not completely cooked.

---

[3]Bardwell and Salunkhe, *Home Drying,* p. 3.

Methods of Vegetable Preparation

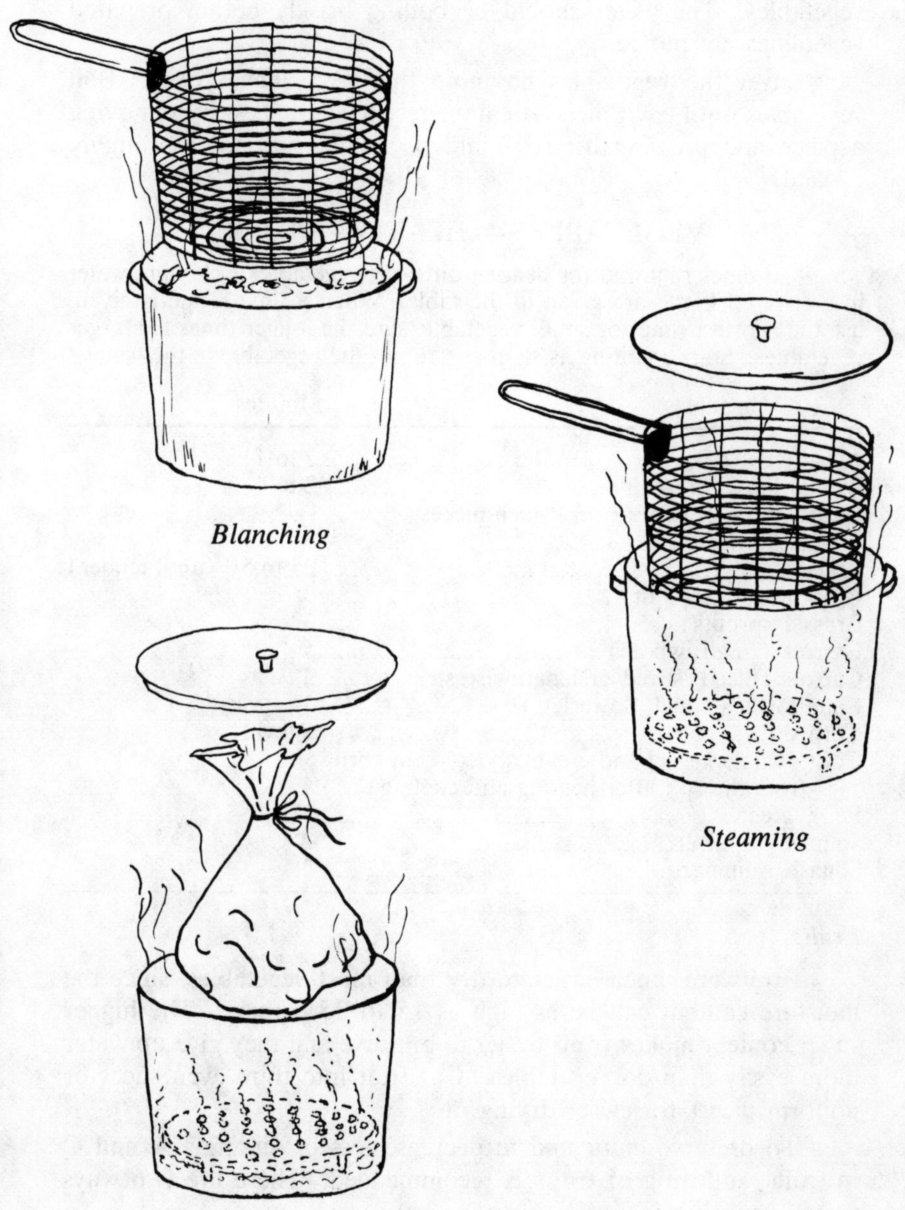

Blanching

Steaming

Steaming in a cloth bag

Blanching. Use a kettle deep enough that when the sieve or basket of vegetables is placed in it, boiling water will cover the vegetables. The water should be boiling briskly before prepared vegetables are put in.

Layer the vegetables no more than 2½ inches deep. Boil vegetables until each piece is completely heated. Test by removing a piece and pressing it. It should be tender but not thoroughly cooked.

VEGETABLES — HEATING TIMES

The times required for heating different vegetables in boiling water in a covered kettle are given in the table. When a time range is given, use the shortest time for small vegetables and the longer times for larger vegetables. Start counting as soon as you put the vegetable in the water.[4]

Vegetables	Minutes
Asparagus stalks	2 to 4
Beans, lima (or pods)	2 to 4
Beans, green or wax (1- or 2-inch pieces or frenched)	3
Beets	25 to 50 (until tender)
Broccoli stalks (split)	3
Brussel sprouts	3 to 5
Carrots (small, whole)	5
Carrots (diced, sliced or lengthwise strips)	2
Cauliflower (1-inch flowerlets)	3
Corn on the cob	7 to 11
Corn (whole kernel and cream style — cut corn from the cob after heating and cooling)	4
Peas, green	1.5
Spinach	2
Squash, summer	3

Fruits

Fruits are much easier to dry than most vegetables, since the moisture content can be as high as 15 to 25 percent. The higher sugar content makes fruits easier to preserve and they give up water more easily than do vegetables. Cut fruit into thin, even slices or uniform pieces for easier drying.

To preserve color and to decrease loss of vitamins A and C in fruits, sulfuring of fruits is recommended. There are two ways to accomplish this.

[4]USDA, *Yearbook of Agriculture,* p. 465.

1. Outdoors. Use a sulfur box. (See the illustration following.)

2. Indoors. Soak fruit in sodium sulfite (do not confuse with sodium sulfate) or sodium bisulfite solution.

Two other, but less efficient, treatments are to dip fruit in salt water bath of four to six tablespoons salt to one gallon of water for about ten minutes, or to precook fruit in steam or boiling water until tender but firm.

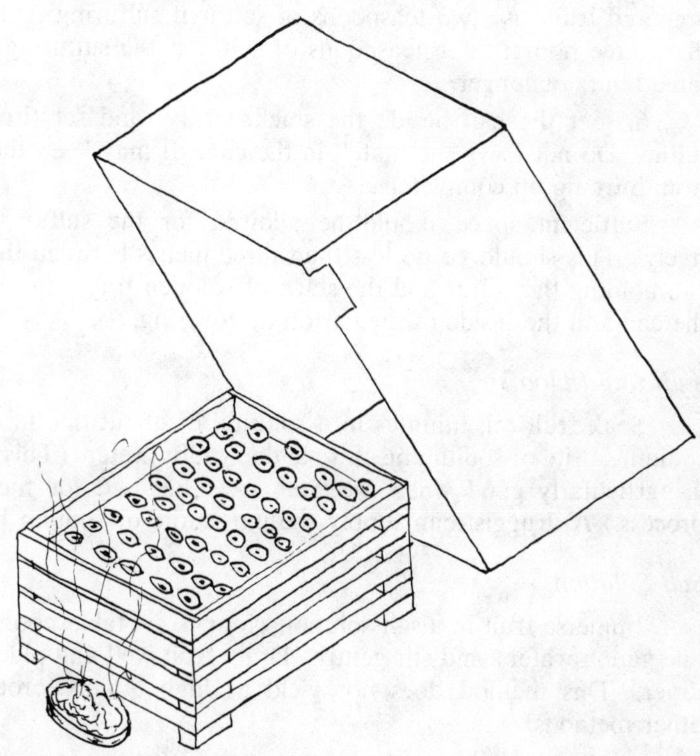

Sulfurer: Preparing fruit for drying

Sulfuring Outdoors

1. Place fruit (skin side down to prevent the loss of juices) on wooden trays having wooden slats, with the fruit not over one layer deep. (Metal will react with the sulfur, so it is important that wooden trays are used.)

2. Stack the trays about 1½ inches apart to permit the sulfur fumes to circulate.

3. Use a tight wooden box or heavy carton to cover the trays. It should be slightly larger than the stacked trays.

4. Cut a small opening at the bottom of the box for lighting the sulfur and for ventilation.

5. Place sulfur in a clean metal container such as a tin can, shallow, but deep enough to prevent overflow. To each pound of prepared fruit, use two teaspoons of sulfur if sulfuring time is less than three hours; three teaspoons of sulfur if the sulfuring time is three hours or longer.

6. Set the can beside the stacked trays and set fire to the sulfur. Do not leave the match in the can. It may keep the sulfur from burning up completely.

Sufficient space should be allowed for the sulfur to burn freely. This should be no less than three inches between the metal can holding the sulfur and the stack of wooden trays, and between the can and the inside of the carton or covering.

Sulfuring Indoors

Soak fruit ten minutes in a solution of about one tablespoon sodium sulfite or sodium bisulfite to one gallon water. (This method is particularly good when the oven is to be used for the drying process.) A druggist can supply sodium sulfite or sodium bisulfite.

Salt Solution

Immerse fruit in a salt solution (four to six tablespoons salt to one gallon water) and stir gently. Drain fruit well and place in the dryer. This method does not yield as high quality products as other methods.

Ascorbic Acid Solution

Fruits may also be immersed and stirred gently in an ascorbic acid solution (one to one and a half tablespoons ascorbic acid to one gallon water) before drying. This preparation retards oxidation, and prevents darkening of light-colored fruits to some extent.

Steaming

Follow method for vegetables. Have only one layer of fruit in basket or steamer.

Drying of Product

After blanching or sulfuring a food, it is ready for the actual drying. There are two main methods of drying, indoor and outdoor.

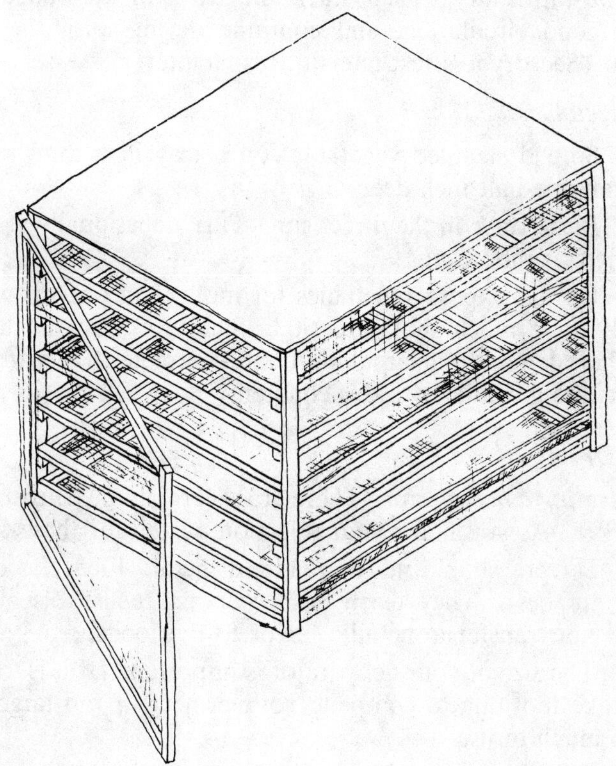

Sun dryer: Used for drying fruits and vegetables

Outdoor Drying

This method can be used only in climates where the sun is very hot and the air is very dry. Partly dried foods should be brought into the garage or carport at night and covered so that they do not absorb moisture from the damp air.

Fruit.

1. Sulfur or treat as directed, then spread the fruit on flat trays, one layer deep. To obtain good circulation at all times, keep the trays separated. This will shorten the drying time.

2. Place the trays in the direct sun. Turn the fruit occasionally. A piece of fine net, cheesecloth or screen should cover each of the trays to protect the fruit from insects.

3. Leave the fruit in direct sun for several days until it is about two-thirds dry. Then stack the trays in the shade, where there is good circulation, and continue drying until the fruit is leathery. (See dryness test later in this chapter.)

Vegetables.

1. Spread steamed vegetables on a tray in a thin layer, not more than one-half inch deep.

2. Place trays in the direct sun. Turn occasionally to prevent scorching or sunburning.

3. Do not expose vegetables for more than one or two days, generally for a much shorter period. Stack the trays in the shade where there is good air circulation and complete the drying. Dryness tests are given later in this chapter.

Indoor Drying

Selecting a dehydrator. There are several home dehydrators on the market. As you make your selection, watch for these features:

1. Screens should not be made of metal. Fiberglas or nylon screens are best. They clean easily, do not cause discoloration, and will not transfer a metallic flavor to the foods.

2. The size of your dehydrator is important. If it is too small, it will take too long to complete your project; if too large, it will cost too much to use.

3. Dehydrators made of metal could be a hazard to small children. They become hot enough (about 150° F.) to cause minor burns.

Drying in a dehydrator.

1. Be sure to follow the manufacturer's specific directions for the dehydrator used.

2. If the dehydrator is used in the house, be sure to have a window or door open to create a crosscurrent. Good air circulation is necessary for drying.

3. Preheat the dehydrator to 140° F., and hold the drying temperature as close to 140° F. as possible. This allows for maximum drying without cooking or too much vitamin loss.

4. Examine the product from time to time. Rotate the trays occasionally to get uniform drying and turn the product if necessary. When a food first starts to dry there is very little chance of scorching, but it scorches easily when nearly dry.

5. The drying time for fruits and vegetables varies with the kind and size of pieces and the load on the tray. Fruits usually take eight to twenty hours to dry properly while vegetables take only six to sixteen hours.

> All products dried in air or sun should be given a final brief heating of 165° to 175° F. before being stored, to kill any microorganisms that may be adhering to the surface. This may be accomplished by spreading the food on trays and reheating it in the oven for 10 to 15 minutes.[5]

Oven drying. Fruits and vegetables can be dried in the oven. To allow for good air circulation, trays should be at least one and a half inches smaller than the inside width and depth of the oven being used.

1. The oven load should not be more than four to six pounds of fruit at one time.
2. Use no more than two to four trays. Place the trays at least two and a half inches apart.
3. Allow a space of three inches at both the top and bottom of the oven.
4. Place a thermometer at the back of the top tray.
5. Maintain a maximum temperature of 140°-150° F.

> Don't turn on the top unit in an electric oven. If necessary, remove it. Turn on the current or light the gas burner 15 minutes before drying time. If there is a regulator, set it at 150°, 200°, or 250° F., whichever is the lowest setting on your oven. If a gas stove has no regulator, turn the flame very low. Be careful throughout drying lest the flame go out unnoticed.
>
> Leave the oven door ajar at least 8 inches when using a gas oven, less if using an electric. Prop an electric oven door open by tucking a folded pot holder in the top corner to make about a half-inch crack. Prop a gas oven door open with an 8-inch stick. The right opening helps control heat and lets out moist air.[6]

---

[5]Bardwell and Salunkhe, *Home Drying,* p. 7.
[6]*Ibid.*

6. To prevent scorching, frequently turn and rotate the trays during drying.

7. Ventilate the room well.

Conditioning

Even if the food has all been dried on the same tray, it may not be uniformly dry. Some pieces may be quite moist when others are "hard dry." Such unevenness is overcome by piling the food in a heap for one or two days longer. It should be stirred occasionally to aid in the uniform distribution of moisture.[7]

Packaging and Storing Dried Foods

Insect damage is a serious hazard in storage of dried vegetables — even more so than in fruit. It is advisable to pack them immediately in proper containers when taken from the dehydrator.

Containers used for packaging should be both moisture-proof and insect-proof and should be sealed with tape or paraffin. Tin cans with tight-fitting lids are excellent. Store containers of any kind in a cool, dry, dark location.

In storage, all dried fruits and vegetables deteriorate in flavor, color, texture, and odor. They should not be stored for more than a year.

Mold will grow only if the moisture content is excessive. This is caused by improper drying. The best protection against mold is thorough drying and proper packaging and sealing.

How to Use

Pour boiling water over the dried fruit in a saucepan, just to cover — no more — and simmer, covered, for ten to fifteen minutes, depending on the fruit and the size of pieces. Remove saucepan from the heat, leaving the cover in place. If sweetening is desired, add it after cooking. Sugar has a tendency to make fruit fibers tough when cooking. Chilling the cooked fruit overnight enhances the flavor. Reconstituted fruits may be used the same as fresh fruits in cooking.

It is not necessary to cook fruits before eating.

Vegetables should always be soaked before cooking. Leave the vegetables in the pan they were soaked in, and do not pour off

[7]Margaret M. Justin, Lucile R. Rust, and Gladys E. Vail, *Foods* (Cambridge, Mass.: Houghton-Mifflin Co., 1948), p. 563.

any excess water that may not have been absorbed. Add only enough water to cover the bottom of the pan, then cover and quickly bring to a boil. Lower the heat and simmer until vegetables are plump and tender.

Fruit Leather

This is a very enjoyable way of preserving fruit. Just how long fruit has been made into a delectable leather is not really known, but Dora D. Flack states in her book *Fun with Fruit Preservation* that "fruit leather was made when people had to crush it in hollowed-out rocks, as they ground grain, and then spread it out in the sun to dry on leather."[8]

Fruit leather can be made whenever you have fruit. During the winter, cookie sheets of leather will dry in front of a sunny window or over a heating unit. An electric heater with a fan to blow warm air across the pans will speed drying. Use your imagination when it comes to finding heat sources.

Sources of fruit for leather.

— Ends and pieces from canning fruit
— Preserves that didn't jell
— Home canned fruit that has darkened, but is still good
— Overripe fruit
— Fruit in season

Preparation.

— Select fruit
— Wash and puree fruit in a blender. A colander or sieve will work also, if you mash the fruit first
— Make sure there are no lumps
— Add sweetening or flavoring if desired (optional)
— Line a cookie sheet with plastic wrap and spread the puree evenly over it, about an eighth of an inch thick. The plastic wrap should extend over the sides and ends of the cookie sheet. A 17″ x 12″ cookie sheet will hold about two cups of puree

---

[8]Dora D. Flack, *Fun With Fruit Preservation* (Bountiful, Utah: Horizon Publishers, 1973), p. 2.

Sun drying.

— Dry puree in direct sunlight nine to ten hours
— Use a home dryer, a screened box or cover it with cheese-cloth
— Leather is dry when the color darkens and it looks and feels like leather
— While fruit is warm, roll it jelly roll fashion, removing the plastic wrap

FRUIT LEATHER — PREPARATION[9]

Product	Preparation	Flavoring
Apples	Puree	Sweeten to taste
Apricots	Puree	Sweeten 1 tablespoon sugar or honey for 1 cup puree (¼ teaspoon cinnamon or dash of nutmeg may be added)
Peaches	Wash and peel, puree	Sweetening optional
Pears	Peel, add small amount of water to help puree fruit if necessary	None necessary
Plums	Wash, pit and puree	Sweeten as desired
Raspberries	Wash and puree	Sweeten as desired
Rhubarb	Add water to first part of fruit	Sweeten as desired

(Lemon Juice: 1 teaspoon to one pint of puree if desired.)

— Place the fruit rolls in a suitable container for storage
— Store in a cool, dry, dark place

Because of variations in sunlight, heat and humidity, the drying time for each fruit will be different.

Oven drying. Follow basic preparation procedures. Set the oven at the lowest possible temperature. After placing trays of

[9]Flora Bardwell, *Fruit Leather* (Logan, Utah: Utah State University Extension Service).

puree in the oven, prop the door open for proper ventilation. It should dry in six to ten hours. Be careful not to overdry. Remove the leather and roll it while it is warm. It becomes difficult to roll once it cools.

Storage. For freezer storage, roll the leather with plastic wrap and seal it so that moisture from the freezer cannot get into it.

To store leather in bottles, roll it without plastic wrap, and place in a clean jar. Place a new lid on and screw the ring on lightly. Place jars in oven preheated to 165° F. and leave twenty to thirty-five minutes. Remove the heated jars and screw the rings down tight. Store in a dry, cool, dark location.

For storage in paper sacks, roll fruit leather with plastic wrap between layers. Leave open at both ends. Store in a paper sack and place in a dry, cool, dark location. Paper boxes can also be used.

For storing in cotton or linen sacks, roll fruit leather (plastic wrap can be used between the layers, but is not necessary) and store in cloth sacks in a dry, cool, dark location.

Problems. Make periodic checks of your fruit leather. If mold grows on it, this could result from several possibilities:

1. The leather was not thoroughly dried before being stored.

2. Moisture is getting into your storage area.

The biggest problem of all will be trying to keep any to store. The secret ingredient of fruit leather seems to be "vanishing cream." It disappears as fast as you can make it.

DRYING TABLE FOR FRUITS AND VEGETABLES

Food	Preparation for Drying	Treatment Before Drying		Maximum Temperature	Approximate Drying Time		Hardness Test (cool a piece before testing)
		Method	Time in Minutes		Dryer	Sun Drying	
Fruit							
Apples	Pare, core and cut in ¼ inch slices or rings. Spread not more than ½ inch deep on trays. Coat slices with an ascorbic acid solution to hold color temporarily.	sulfur steam blanch	60 min. 5 min.	Begin at 130° F., raise to 150° F. after 1 hour, reduce to 140° F. when nearly dry	6 hrs.	3 days bright sun	Pliable springy feel; creamy white, no moist area in center
Apricots	Peel if desired. Cut in half, remove pits. Dry pit side up.	sulfur steam	If peeled 30 min. If unpeeled 2-3 hrs. If quarter 2 hrs. If sliced 1 hr. 5-7 min.	Begin at 130° F., raise to 145° F. after 1 hour, reduce to 140° F. when nearly dry	halves — 14 hrs. slices — 6 hrs.	3 days bright sun	Pliable and leathery
Berries	Pick over, wash if necessary. Check skins of blueberries, huckleberries, currants and cranberries. Soft berries such as strawberries do not dry well. They are especially bland and unrecognizable when dried.	dip in boiling water, plunge into cold water	15-30 sec.	Begin at 120° F., raise to 130° F. after 1 hour, raise to 140° F. when nearly dry	4 hrs.	18-20 hrs. bright sun	Hard. No visible moisture when crushed
Cherries	Remove stems. Pit only large cherries. Stems from small cherries may be removed more easily after drying.	dip in boiling water, cool immediately White cherries may be sulfured	15-30 sec. 10-15 min.	Begin at 130° F., raise to 145° F. after 1 hour, reduce to 125° F. during last hour	6 hrs.	18-20 hrs. bright sun	Leathery but sticky

DRYING TABLE FOR FRUITS AND VEGETABLES — (Continued)

Food	Preparation for Drying	Treatment Before Drying		Maximum Temperature	Approximate Drying Time		Hardness Test (cool a piece before testing)
		Method	Time in Minutes		Dryer	Sun Drying	
Figs	If figs are small or have partly dried on the tree, they may be dried whole without blanching. If figs are large, cut in half.	steam	20 min.	Begin at 120° F., raise to 145° F. after 1 hour, reduce to 135° F. when nearly dry	5 hrs. for halves	20-24 hrs. bright sun	Glossy skin, slightly sticky, pliable, leathery
		dip in boiling water	15-30 sec.				
Grapes	Use Thompson or seedless varieties for drying only. Leave whole, remove stems. Check skins.	dip in boiling water	15-30 sec.	Begin at 120° F., raise to 145° F. after 1 hour, reduce to 135° F. during last hours	8 hrs.	18-20 hrs. bright sun	Pliable, leathery, dark brown
Nectarines	Peel if desired. Cut in half, remove pits. Dry pit side up.	sulfur	If peeled 20 min. If unpeeled — 2-3 hrs. If quartered — 2 hrs. If sliced — 1 hr.	Begin at 130° F., raise to 145° F. after 1 hour, reduce to 140° F. when nearly dry	halves — 14 hrs. slices — 6 hrs.	5 days bright sun	Pliable and leathery
		steam	5-20 min.				
Peaches	Peel if desired. Cut in half, remove pits. Dry pit side up. Sulfur outdoors.	sulfur	If peeled 30 min. If unpeeled — 2-3 hrs. If sliced 1 hr.	Begin at 130° F., raise to 145° F. after 1 hour, reduce to 140° F. when nearly dry	halves — 14 hrs. slices — 6 hrs.	5 days bright sun	Pliable and leathery
		steam	5-20 min.				

DRYING TABLE FOR FRUITS AND VEGETABLES — (Continued)

Food	Preparation for Drying	Treatment Before Drying		Maximum Temperature	Approximate Drying Time		Hardness Test (cool a piece before testing)
		Method	Time in Minutes		Dryer	Sun Drying	
Pears	Pare and remove core and woody vein. Leave in halves or slice. If slicing pare off skin. To prevent oxidizing, coat with ascorbic acid.	sulfur outdoors dip in solution or precook steam blanch	2-4 hrs. slices — 5 min. halves — 20 min.	Begin at 130° F., raise to 145° F. after 1 hour, reduce to 140° F. when nearly dry	halves — 15 hrs. slices — 6 hrs.	2-4 days bright sun	Leathery, springy feel No moisture when cut and squeezed
Plums	Cut in halves and remove pits or leave whole. Soften and crack skins.	sulfur steam blanch in boiling water, dip in lye broth (3 T lye to 1 gal. water)	20-25 min. 2 min. 2 min. 30 sec.	Begin at 120° F., raise to 145° F. after 1 hour, reduce to 140° F. when nearly dry	whole — 14 hrs. halves — 8 hrs. slices — 6 hrs.	18-20 hrs. bright sun	Pliable and leathery
Prunes	Cut in halves and remove pits or leave whole. Halves; no pre-treatment. Whole — crack skins.	steam blanch dip in lye bath (3 T lye to 1 gal. water)	2 min. 2 min. 30 sec.	Begin at 120° F., raise to 145° F. after 1 hour, reduce to 140° F. when nearly dry	whole — 14 hrs. halves — 8 hrs. slices — 6 hrs.	18-20 hrs.	Pliable and leathery
Vegetables							
Asparagus	Use 3-inch tips only, split lengthwise after cooking.	steam	10 min. or until tender but firm	Begin at 130° F., raise to 145° F. after 1 hour, reduce to 140° F. when nearly dry	10-12 hrs.		Very brittle, greenish black

DRYING TABLE FOR FRUITS AND VEGETABLES — (Continued)

Food	Preparation for Drying	Treatment Before Drying		Maximum Temperature	Approximate Drying Time		Hardness Test (cool a piece before testing)
		Method	Time in Minutes		Dryer	Sun Drying	
Beans, snap	Trim and slice lengthwise or cut in 1-inch pieces.	steam	15-20 min. or until tender but firm	Begin at 120° F., raise to 145° F. after 1 hour, reduce to 130° F. when nearly dry	8-10 hrs.		Brittle, dark green to brownish
Beans — green lima	Shell	steam	15-20 min. or until tender but firm	Begin at 135° F., raise to 145° F. after 1 hour, reduce to 130° F. when nearly dry	8-10 hrs.		Hard, brittle; shatters when hit with hammer
Beans, soy	Blanch pods and shell.	blanch pods	5-7 min.	Begin at 130° F., raise to 140° F. after 1 hour, reduce to 135° F. when nearly dry	6-8 hrs.		Shatters when hit with a hammer
Beets	Select small tender beets of good color and flavor. Trim off all but 1 inch of crown and root. After steaming remove crown and root and peel. Slice ⅛ inch thick. May be diced no larger than ¼ inch square. Spread not more than ¼ inch deep on trays.	steam	30-60 min.	Begin at 120° F., raise to 150° F. after 1 hour, reduce to 130° F. when nearly dry	14-18 hrs.		Tough, brittle dark red

DRYING TABLE FOR FRUITS AND VEGETABLES — (Continued)

Food	Preparation for Drying	Treatment Before Drying		Maximum Temperature	Approximate Drying Time		Hardness Test (cool a piece before testing)
		Method	Time in Minutes		Dryer	Sun Drying	
Broccoli	Trim, slice lengthwise in ½ inch strips. Wash.	steam	8-10 min.	Begin at 120° F., raise to 145° F. after 1 hour, reduce to 130° F. when nearly dry	9-11 hrs.		Brittle, very dark green
Brussels Sprouts	Cut lengthwise in ½ inch thick strips.	steam until tender	12 min.	Begin at 120° F., raise to 145° F. after 1 hour, reduce to 130° F. when nearly dry	8-10 hrs.		Crisp, dark green
Cabbage	Remove outer leaves. Core. Cut in strips ¼ inch thick. Spread evenly on tray not more than ½ inch deep. Occasionally lift and turn food on the tray to prevent matting.	steam	5-10 min. or until tender	Begin at 120° F., raise to 140° F. after 1 hour, reduce to 130° F. when nearly dry	8-10 hrs.		Tough to brittle crisp. Pale yellow to green
Carrots	Select crisp, tender carrots, free from woodiness. Wash. After steaming trim off roots and top. Scrape or peel. Cut into slices or strips not more than ⅛ inch thick.	steam	8-10 min.	Begin at 120° F., raise to 150° F. after 1 hour, reduce to 130° F. when nearly dry	14-18 hrs.		Tough, very brittle, deep orange
Cauliflower	Separate into flowerlets, cut large ones in half.	dip in salt solution (6 T of salt to 1 gal. of water) then steam	10 min. or until tender but firm	Begin at 120° F., raise to 145° F. after 1 hour, reduce to 130° F. when nearly dry	9-11 hrs.		Hard to crisp tannish yellow

DRYING TABLE FOR FRUITS AND VEGETABLES — (Continued)

Food	Preparation for Drying	Treatment Before Drying		Maximum Temperature	Approximate Drying Time		Hardness Test (cool a piece before testing)
		Method	Time in Minutes		Dryer	Sun Drying	
Celery	Strip off leaves, cut stalks into ½ inch pieces. Stir occasionally during drying.	steam	10 min.	Begin at 130° F., raise to 150° F. after 1 hour, reduce to 130° F. when nearly dry	9-10 hrs.		Dry, brittle, shatters when hit with a hammer
Corn (cut)	Husk, trim. Steam on cob until the milk is set. Cut from cob. Spread ½ inch deep on tray.	steam	15 min.	Begin at 130° F., raise to 145° F. after 1 hour, reduce to 130° F. when nearly dry	8-10 hrs.		Dry, brittle shatters when hit with a hammer
Corn-on-the-cob (pop corn)	Allow to mature in the field and to become partially dry in the husk on the stalk.	air dry	until dry to the feel				Rub easily from cob when kernels are dry
Corn, cream	Husk, trim, steam on cob until the milk is set. Cut from cob. Add 1 cup cream or rich milk, ¼ cup sugar, 2 T salt in each gallon cut corn. Spread on trays not more than ¼ inch deep.	steam simmer on stove until cream is absorbed, stir constantly	15 min.	Begin at 120° F., raise to 140° F. after 1 hour, reduce to 130° F. when nearly dry	12-14 hrs.		Dry, brittle, shatters when hit with a hammer
Eggplant	Peel and slice ⅛ to ¼ inch thick. Dip immediately in a solution of 6 T vinegar to 1 gallon water for 25 minutes. Steam at once.	dip and steam	5-10 min. or until tender	Begin at 120° F., raise to 140° F. after 1 hour, reduce to 130° F. when nearly dry	8-10 hrs.		Crisp, very dark green

DRYING TABLE FOR FRUITS AND VEGETABLES — (Continued)

Food	Preparation for Drying	Treatment Before Drying		Maximum Temperature	Approximate Drying Time		Hardness Test (cool a piece before testing)
		Method	Time in Minutes		Dryer	Sun Drying	
Garlic	Very seldom dried at home. However, if you desire to dry it, treat it like onions.						
Greens	Trim off tough, thick stems. Spread leaves about ¼ inch deep on tray.	steam	5-20 min. or until tender	Begin at 120° F., raise to 140° F. after 1 hour, reduce to 130° F. when nearly dry	8-10 hrs.		Crisp, very dark green
Herbs	Includes celery leaves as well as the greenery from aromatic herbs — basil, parsley, sage, tarragon, etc. Use only the most tender and flavorful leaves from the upper 6 inches of the stalk.	all are air dried and should be kept in the dark	until dry				Crisp — color deeper
Mixed Vegetables	Mixed vegetables are never dried in combination; drying times and temperatures vary too much between the various vegetables. Dry the vegetables and season them separately, then combine them in packages according to your desire.						

DRYING TABLE FOR FRUITS AND VEGETABLES — (Continued)

Food	Preparation for Drying	Treatment Before Drying		Maximum Temperature	Approximate Drying Time		Hardness Test (cool a piece before testing)
		Method	Time in Minutes		Dryer	Sun Drying	
Mush-rooms	Use only fresh, unbruised, young mushrooms. Dry whole or sliced, depending on size.	no pre-cooking necessary		Begin at 130° F., raise to 145° F. after 1 hour, reduce to 140° F. when nearly dry	8-10 hrs.		Leathery to brittle
Okra	Use young, tender pods only. Cut ½ inch cross-wise slices and split length-wise. Spread not more than ½ inch deep on trays.	steam	5-8 min.	Begin at 130° F., raise to 145° F. after 1 hour, reduce to 130° F. when nearly dry	8-10 hrs.		Very brittle
Onions	Remove outer discolored layers. Slice ⅛ inch thick. Keep slices uniform.	steaming not necessary		Begin at 140° F., reduce to 130° F. when nearly dry	8-10 hrs.		Brittle, light-colored
Parsnips	Same as carrots.	steam	8-10 min.	Begin at 120° F., raise to 150° F. after 1 hour, reduce to 130° F. when nearly dry	8-10 hrs.		Very brittle
Peas, black-eyed	Treat like beans (shell) above.						Hard, very brittle
Peas, green	Select young, tender peas of a sweet variety. Shell. Stir frequently during the first few hours of drying.	steam imme-diately	15 min. or until tender	Begin at 135° F., raise to 145° F. after 1 hour, reduce to 130° F. when nearly dry	8-10 hrs.		Wrinkled, hard shatters when hit with a hammer

DRYING TABLE FOR FRUITS AND VEGETABLES — (Continued)

Food	Preparation for Drying	Treatment Before Drying		Maximum Temperature	Approximate Drying Time		Hardness Test (cool a piece before testing)
		Method	Time in Minutes		Dryer	Sun Drying	
Peppers, hot (chili)	Select mature, dark red pods. String them on a string and hang them on a south wall. Pods will shrink and be dark when dry.	no pre-cooking necessary					Pliable, can be bent without snapping
Peppers and pimentos	Cut into ½ inch strips or rings. Remove seeds. Spread rings two layers deep. Spread strips not more than ½ inch deep.	steam	10 min.	Begin at 120° F., raise to 145° F. after 1 hour, reduce to 135° F. when nearly dry	8-10 hrs.		Pliable
Potatoes, sweet (and yams)	Use only firm, smooth sweet potatoes and yams. After steaming, trim, peel, cut into ⅛ inch slices, or shred.	rinse in cold water, steam	4-6 min.	Begin at 120° F., raise to 150° F. after 1 hour, reduce to 130° F. when nearly dry	14-18 hrs.		Brittle
Pumpkin	Quarter, remove seeds and pith, cut in 1 inch strips and peel. May be shredded.	steam	8-13 min.	Begin at 120° F., raise to 145° F. after 1 hour, reduce to 130° F. when nearly dry	14-18 hrs.		Leathery, chips may be brittle
Rhubarb	Cut in 1 inch lengths.	dip in boiling water	3 min.	Begin at 130° F., raise to 145° F. after 1 hour, reduce to 130° F. when nearly dry	8-10 hrs.		Very brittle dark green and red

DRYING TABLE FOR FRUITS AND VEGETABLES — (Continued)

Food	Preparation for Drying	Treatment Before Drying		Maximum Temperature	Approximate Drying Time		Hardness Test (cool a piece before testing)
		Method	Time in Minutes		Dryer	Sun Drying	
Rutabagas	Quarter and peel, cut into ⅛ inch wide slices or strips.		15 min. or until tender but firm	Begin at 130° F., raise to 145° F. after 1 hour, reduce to 130° F. when nearly dry	8-10 hrs.		Leathery
Spinach and other greens	Choose young tender leaves. Wash. Make sure that leaves are not wadded when placed on trays. Cut large leaves crosswise into several pieces. Try not to overlap leaves on tray.	steam	4-6 min.	Begin at 140° F., raise to 145° F. after 1 hour, reduce to 140° F. when nearly dry	8-10 hrs.		Brittle, loosely crumbled
Squash, banana	Wash, peel and slice in strips ¼ inch wide.	steam	8-13 min.	Begin at 120° F., raise to 145° F. after 1 hour, reduce to 120° F. when nearly dry	14-18 hrs.		Tough to brittle
Squash (Hubbard) yellow	Treat the same as pumpkin.	steam	until tender	same as for pumpkin	14-18 hrs.		Tough to brittle
Squash (summer) crookneck, scallop zucchini, etc.	Wash, trim and cut into ¼ inch slices.	steam	8-13 min.	Begin at 120° F., raise to 145° F. after 1 hour, reduce to 135° F. when nearly dry	14-18 hrs.		Brittle

DRYING TABLE FOR FRUITS AND VEGETABLES — (Continued)

Food	Preparation for Drying	Treatment Before Drying		Maximum Temperature	Approximate Drying Time		Hardness Test (cool a piece before testing)
		Method	Time in Minutes		Dryer	Sun Drying	
Tomatoes	Select tomatoes of good color. Dip in boiling water. Chill immediately by plunging into cold water. Peel, remove stem end, slice ⅛ inch thick.	steam or dip in boiling water	1 min.	Begin at 125° F., raise to 145° F. after 1 hour, reduce to 130° F. when nearly dry	14-16 hrs.		Leathery, dull red
Turnips	Very seldom dried. However, if you wish to dry them, quarter and peel. Slice in ⅛ inch slices or shred.	steam	15 min. or until tender	Begin at 120° F., raise to 150° F. after 1 hour, reduce to 130° F. when nearly dry	14-18 hrs.		Leathery, brittle chips
Powdered Vegetables	"For use in soup or puree, powder leafy vegetables after drying by grinding fine in a blender or an osterizer."*						
Soup Mixture	"Cut vegetables into small pieces; prepare and dry according to directions for each vegetable. Combine and store. Satisfactory combinations may be made from cabbage, carrots, celery, onions and peas. Rice, dry beans, or split peas, and meat stock are usually added at the time of cooking."**						

*Bardwell and Salunkhe, *Home Drying*, p. 10.
***Ibid.*

Note:

1. Spread in single layers on trays.
2. Usual drying temperatures of 140° F. - 150°F. Onions and cabbage require temperature not above 135° F.
3. Fruits dry more rapidly if cut in slices or quartered.

11
Brining

Brining, preserving vegetables by using salt, has been done for many years. Before it was known how to can foods so that they would keep, a great deal of the winter's food supply was preserved in this way. In many rural areas, brining is still used as a method of food preservation. It provides a distinctive taste. If you haven't tasted food prepared this way, try a little. You may need this method. For instance, if the power was off for a long time, you could salvage a good deal of the food in a deep freeze by brining, rather than lose it all.

Salt does two things to the vegetables. First, it causes the withdrawal of water and some solids from the vegetable, and second, it acts as an antiseptic and prevents spoiling.

An important factor in brining vegetables is the amount of sugar vegetables contain. The more natural sugar present in the vegetable, the milder the brine solution required. Cabbage, for instance, contains considerable sugar, so a heavy brine is not needed.

Many vegetables may be preserved in salt or strong brine without causing any marked changes in flavor or texture.

Basically, there are two ways of preserving foods by brining: the dry salt method and the wet salt method. Neither is difficult, though both are time-consuming.

Dry Salt Method

With the dry salt method, the salt draws moisture out of the vegetable to form a brine. Cabbage, green beans, sliced beets, greens of all sorts, turnips, lettuce, sliced root vegetables — all are suitable for this process.

1. Vegetables need not be blanched.

2. Fill stone jar or enamel container with alternate layers of vegetables and salt to within three inches of the top. Use half a pound of salt (pickling salt must be used) to each ten pounds of vegetables. Be sure the salt is evenly distributed.

3. Cover with a plate or disc, then with cheesecloth or similar clean cloth, and weight down with a heavy object. (See heading below, "Equipment Needed.")

4. If not enough brine forms to completely cover the vegetables, add additional brine solution of one pound of pickling salt to two quarts of water.

5. Let stand until fermentation is completed. Remove scum from time to time throughout the process.

6. Pack in clean sterilized jars. Pack very full and compact. Work a knife blade through the jar to make sure all the air is out. Use a damp, clean cloth to wipe off the rim and edges. Prepare lids by placing them in pan of hot water. Put on hot lid and screw the ring on tightly.

Vegetables processed in this way keep a long time.

Wet Salt Method

With the wet salt method a brine is made of pure salt and water and poured over the food. This method is suitable for vegetables such as corn, green beans, peppers, cauliflower, peas, and green lima beans.

1. Blanch all vegetables according to the chart in chapter 9 on freezing foods.

2. Fill a stone jar or enamel container to within three inches of the top with vegetables.

3. Make a brine of three and a half to four pounds of salt per gallon of water. Pour cool brine over the vegetables until they are completely covered.

4. Cover the food with a plate or disc, then a clean cloth, and weight it down with a heavy object. (See heading below, "Equipment Needed.")

5. Allow vegetables to stand until fermentation is complete. Remove the scum several times during the processing.

6. Pack in clean, sterilized jars. Pack very full and compact. Work a knife blade through the jar to make sure all the air is out. Use a damp, clean cloth to wipe off the rim and edges. Put on a hot lid and screw the ring on tightly.

Care of Foods During Fermentation

A scum will form on the top. "This scum is usually composed of wild yeasts, molds and bacterias. This scum must be removed carefully at frequent intervals because it eats the acid and then spoilage occurs."[1] Every speck of scum must be removed before the vegetables are sealed.

Ten to sixteen days are usually required to complete the brining process. Time will vary according to the vegetable, the strength of the brine and the room temperature.

A simple way to tell if fermentation is complete is to give the jar or container a little shake. If you cannot see or hear any bubbles come to the surface, the process is complete and the vegetables are ready to eat or be packed and sealed. If the fermenting process is not complete before sealing, a gas could form within the sealed container and cause spoilage.

Rules for Success

Be sure to use *pickling salt* for all brining. The use of table salt is not advised. Be sure the salt is not lumpy; it must be evenly distributed.

Be sure the vegetables are submerged in the brine. If they are allowed to float above the level of the brine, spoilage will occur.

Use no metal equipment in brining where it will contact the brine.

Keep the scum controlled. Skim it off at regular intervals or spoilage will occur.

Remove the cheesecloth, plate and weights when taking off the scum. Thoroughly wash the cloth, plate and weights before replacing them.

To be sure that fermentation is complete, allow the vegetables to stand one or two days after they are apparently ready to pack.

---

[1]*Home Guide to Food Preservation* (Westinghouse Electrical and Mfg. Co.), p. 42.

Brining crock

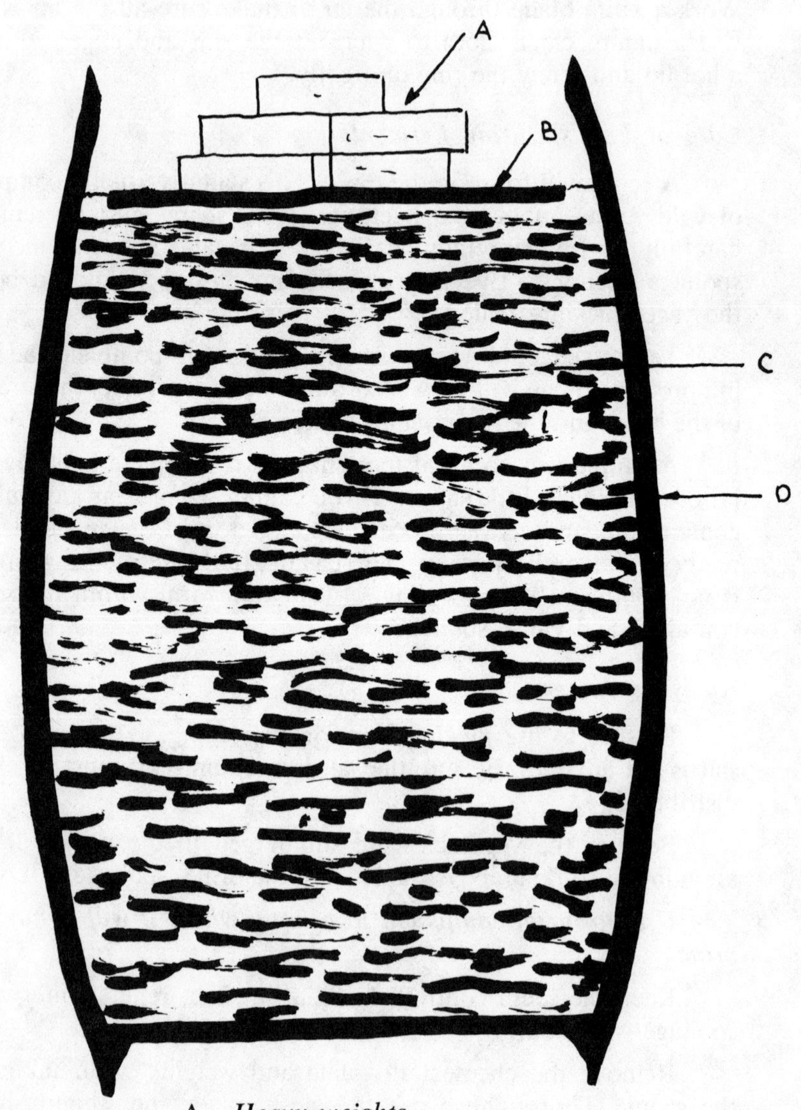

A. *Heavy weights*
B. *Large plate or clean piece of wood*
C. *Prepared vegetables, salt and brine*
D. *Large glass jar or suitable container*

Equipment Needed

You will need the following equipment for brining:

1. A large crock, stoneware jar, enamel container, or gallon jars.

2. A large crockery or china plate that will fit upside down inside the top of the container. If you do not have a plate, a large clean piece of wood may be used.

3. A heavy weight, such as several bricks or large rocks which have been cleaned. If no bricks or rocks are available, use one or more glass jars filled with either sand or water.

4. Clean cloth or cheesecloth. (This should keep out any insects.)

5. Canning jars, lids, and rings.

6. Utensils you would normally use in food preparation, such as knives, spoons, bowls, etc.

How to Use Brined Foods

Sauerkraut made from cabbage need not be washed, but all other vegetables should be or they will be too salty to eat.

To each cup of vegetables, add one quart of hot water. Allow the vegetables to stand for one hour in the hot water, then drain. Cover once more with the same amount of water and rinse thoroughly.

Vitamin Content

A problem about brining is that it removes most of the water-soluble vitamins. Other vitamins remain.

Brining may not be the best way of preserving foods, but it is a way of doing so without refrigeration, freezers or root cellars. You might try it — before you need to know how. You might like the taste.

12
Home Gardens

To some, gardening brings visions of weeds and work. To others, it brings the total satisfaction of being close to earth and growing things. Many people, blessed with a "green thumb," can grow anything. My talent seems to be growing weeds and rocks. In fact, our rocks actually seem to multiply, so naturally ours is a rock garden.

The goal of this chapter is not to debate the merits of various kinds of fertilizer or to give step-by-step instructions on how to plant and grow a garden. Rather, it is to persuade you that no matter where you dwell — city, suburbs, or country — a garden is both practical and possible, but most of all, exciting and fun.

By raising a garden, you grow some of your own food, thus combating the increasingly high cost of living. And home-grown produce tastes so much better than "store bought." In some parts of the country you can grow food year round.

Any surplus food from the garden can be frozen or canned. What a treat in the winter to bring out a favorite vegetable, not only home canned but home grown!

Tools

Few tools are required — a spade or digging fork, a hoe, and a rake. Some other tools are helpful, but unless they fit into the budget, you can have a garden without them. If you need tools, check the second-hand stores and garage sales. Larger, heavy-duty equipment such as rototillers can be rented. You might share the rental cost with a neighbor.

Where Can a Garden Grow?

Traditionally, a garden plot has been envisioned as a square of furrowed soil. This is marvelous if you happen to have such a piece of ground, but a little garden can spring up in the most unexpected places. All it takes is a little thought, effort, faith, and the perseverance to keep ahead of the weeds. Of course, space is the dominating factor in the size and scope of your garden.

No Yard at All?

If you are an apartment dweller, please don't skip to the next chapter, saying, "There's no hope for me here." Just turn on your imagination and let it work for you.

If your apartment has a small patio or balcony, the problem is solved. Look for a large sunny window sill. Does your building have a flat roof that you have access to? Consider a mini-garden. It doesn't have to be all in one place. Consider various containers. Are you excited yet?

A Small-Container Garden

Naturally you can't grow a bumper crop of corn in small places, but there are many things you can grow.

For containers, try boxes, plastic tubs, clay pots, wooden boxes, half barrels, planter boxes made from scrap lumber, plastic dish pans, old bushel baskets, washtubs, flower pots, trash cans; in short, anything that holds dirt will do. Just be sure that the container is large enough to hold the plant when it is fully grown.

Cherry tomatoes and dwarf yellow tomatoes grow well in flower pots.

If you have a railing around a balcony, plant peas, beans or cucumbers in buckets or tubs and let the vine "climb" the railing.

Bush zucchini squash or yellow crooknecked squash can be grown in a tub or bucket; radishes and green onions in a shallow, flat box; tomatoes in buckets, boxes or tubs. Stake up the tomato plant before the plants are too big, so roots are not damaged.

Green peppers do well in flower pots, and herbs such as chives, parsley and sage also will thrive in a container garden.

Hanging pots do not necessarily have to have fancy flowers in them. Plant them with vegetables for a change.

A small amount of chard will grow in a bucket or tub. You would harvest it by cutting the outside leaves, and this would provide a whole season's worth of greens. The plant continues to produce.

If you have the space for several boxes, you could grow lettuce, beets, or carrots — and definitely radishes.

Containers and Dirt

There should always be drainage holes along the sides at the bottom of the containers.

Lightweight gardening soil available in garden shops is a commercial planter mix that should give good results for small areas. Alternatively, any good rich soil will do.

Watering of Container Plants

Proper watering of plants in containers is very important. Since the amount of soil is small, be careful not to overwater. Too much water can cause the soil to sour, and the plant will then die. When the soil becomes dry for a quarter of an inch down, it is time to water.

If you live in an upstairs apartment, remember that water runs, drips and leaks. Be considerate and don't saturate the plants to the point where the water might cause a problem for those downstairs. Heavy plastic or a tray of some sort under the containers would prevent leakage.

Vacant Lot

Look for a vacant lot. Many landowners are allowing empty lots to be gardened. Get together with your neighbors and see if you can work a "city garden."

Just a Tiny Yard

Even if your yard area is very small, there are still ways you can have a garden.

No rule says that all the garden must be grown in one area. Try planting in small patches. That is how the Salsbury garden has to be arranged. In small areas around the trees and clothesline poles, I planted a few herbs. One side of the flower bed has raspberry bushes for shrubs. In another small patch is planted the bell peppers. Cucumbers are climbing the fence. Large squash and melons surround the roof of the root cellar where the vines can grow over the top and use the space very well. Nothing else could be planted there, since the soil is very shallow.

Make use of barrels or planter boxes along the edges of driveways and patios. They look nice and can grow quite a bit of produce.

Border all of your flower beds with vegetables such as carrots, beets, radishes.

Plants that climb, such as peas, beans, cucumbers, and squash, can be placed along fences. Tomatoes can climb a fence rather than being staked in rows.

To thoroughly use a small amount of space, try companion planting. Corn, peas or beans, and squash can all go in the same area. The corn grows high; the peas can climb the corn stalks; and the squash can grow in the spaces in between.

Many bush varieties of peas, beans and squash now available do not require the space or poles of the climbing varieties. On the other hand, the climbing varieties may be exactly what you need. With a trellis or fence your garden can grow up instead of out. Pole tomatoes, cucumbers, beans, peas, and small squash will grow up. With a soil area only two feet wide and eight feet long you could grow quite a bit — *up*.

Paper Garden First

For best results, it's wise to plant a "paper garden" first. Take time to sit down and realistically plan what you can grow. Potatoes and corn may be out for an apartment dweller, but tomatoes and green peppers can be on everyone's list. Once you start planning it is amazing how much "acreage" can be found in a side yard.

What Are the Easy Plants?

To insure success in gardening, especially the first time, choose easy plants. Some fruits and vegetables will grow practically by themselves (weeds don't count); others need to be coddled. Following is a list of fruits and vegetables which are easily grown and produce an abundant yield:

Beans — green or wax	Lettuce	Herbs
Beets	Onions	Grapes
Broccoli	Peas	Melons
Carrots	Peppers	Dwarf fruit trees
Chard	Radishes	Strawberries
Corn	Squash	Raspberries
Cucumbers	Tomatoes	

Specific planting instructions are given on seed packets.

Watering

Watering is an important phase of gardening. How much is as important as how often.

Most vegetables are about 80 percent water, and must have an even supply of moisture — one to two inches per week. In some areas, rainfall will take care of this. If irrigation is necessary, it's best to water deeply and less frequently, as shallow watering encourages tiny roots to remain near the surface. Deep watering allows the plants to develop a strong root system.

> You can overwater. Excess water, especially around plant roots in clay soils, shuts off oxygen, suffocates the plants, and they die.
>
> Heavy soakings in sandy soils wash out nitrogen and cause pale color of foliage. Most plants will be well satisfied with soil moistened to a depth of a foot. Sandy soils dry out faster than clay, and you can tell by looking at your plants how often you should apply the water. Any water that runs off is wasted.[1]

Planting and Thinning

Most beginning gardeners get carried away with enthusiasm and tend to overplant. Crowding vegetables is as bad as allowing weeds to take over. You cannot have a good crop if the plants have no "elbow" room to grow in. Here's where I have my worst problem in gardening. Because I can hardly bear to pull out tiny plants that are doing well, my husband says I will never make a "proper" farmer.

If you become overzealous in seeding, you are only wasting money, so seed sparingly.

---

[1]George Abraham, *The Green Thumb Book of Vegetable Gardening* (New York: The Dell Publishing Co., 1973), p. 7.

Another way to plant is with the pony packs of plants already started. This is more expensive than buying seeds, but is still far less expensive than buying all of the produce at the store. It eliminates the problems of seeding and thinning, and has the advantage that the plants are already established.

A Compost Heap

A compost heap is one of the most economical ways to renew soil nutrients. It can be as elaborate or simple as you desire. Compost can be made of leaves or any other organic matter available, such as lawn clippings, sawdust, garbage, manure, pruning clippings. Pile them in an area away from the house as you collect them. (If you live in an apartment or condominium, try using a large covered garbage can as a compost pile for your container gardens.) Build your pile in thin layers, three to six inches thick. From time to time cover it with about one inch of soil and then soak it with water. Sprinkle it with lime once in a while. Stir the heap occasionally to admit air. The top should be kept concave to collect rainfall.

Compost usually takes a few months to a year to decay, depending on how large and how deep the pile is.

Use compost as you would fertilizer, to prepare soil for planting and to apply later to feed the soil.

Mulching

With mulching, the days of hoeing and weeding are almost gone. To grow, weeds need sunshine, so if you take it away they don't grow. Straw, grass, hay and chopped-up leaves work well as mulch. Mulch also saves moisture and helps to build up the soil if it is turned into the ground later.

Weeds

Some may question all the effort put forth to get ahead in the never-ending battle with weeds.

> Weeds use water and nutrients, carry disease, harbor insects and interfere with the growth of crops. The best time to contest weeds is when they are small, one to three inches high. At this

stage of development the average home garden can be weeded in minutes. Left to grow 12-24 inches high, the same area would require hours to cut the same weeds.[2]

Whatever you do, don't allow weeds to go to seed in the garden area. One plant can populate an entire yard with weeds.

Most of all, enjoy your garden. It's such fun to dig in the dirt and breathe the smells of the earth, such a thrill to see tiny shoots peek through the soil and develop into plants. Sometimes the best psychological treatment in the world is to take out your frustrations on the weeds and allow the peace of the earth to seep into your soul.

GUIDE FOR PLANTING

Very Hardy	Hardy	Tender	Very Tender	For Fall Harvest
Plant as soon as the ground can be worked in the spring	Plant at time of the average last killing frost	Plant two weeks after average date of last killing frost	Plant when the soil and weather are warm	Plant in late June or early July
Broccoli — plants	Carrots	Beans	Beans — lima	Beets
Cabbage — plants	Cauliflower — plants	Corn	Cantaloup	Broccoli
Endive	Beets	Potatoes — early	Celery — plants	Cabbage
Kohlrabi	Onion seed	Tomato — seed	Crenshaw melons	Cauliflower
Lettuce	Parsnips		Cucumbers	Kohlrabi
Onion sets	Swiss Chard		Eggplant — plants	Lettuce
Parsley			Pepper — plants	Radishes
Peas			Pumpkins	Spinach
Radishes			Potatoes for winter	Turnips
Spinach			Squash	
Turnips			Tomato — plants	
			Watermelons	

Vegetable Garden Guide, Utah State University Extension Service No. EC 337.

Note: Spring comes late in areas of high elevation. Late summer planting is not advised in areas where the growing season is very short.

[2]Abraham, *Green Thumb Book,* p. 9.

HOW DEEP TO PLANT

½ inch deep	1 inch deep	2 inches deep
Broccoli	Beets	Beans
Cabbage	Eggplant	Cantaloup
Carrots	Okra	Corn
Cauliflower	Pepper	Crenshaw
Celery	Radishes	Cucumbers
Endive	Salsify	Peas
Kale	Spinach	Pumpkin
Kohlrabi	Swiss Chard	Squash
Lettuce	Tomato	Watermelons
Onions		
Parsley		
Parsnips		
Rutabaga		
Turnips		

Vegetable Garden Guide, Utah State University Extension Service No. EC 337.

For more specific directions check individual seed packets or consult your local extension service.

13
Root Cellars

Root cellaring, wintering over, common storage, putting food back — all these are old-fashioned terms for keeping fruits and vegetables fresh and edible from fall harvest until spring.

Root cellar storage undoubtedly goes back to the time when man first began to gather food in an attempt to provide for the winter season. Farm dwellers in many sections of the country, and even people in small cities, find the root cellar a way of providing storage for garden produce. As a means of food preservation the root cellar is not as effective as freezing, canning or some of the other methods. It is not to be considered for long-term storage, but is one more way that you can use to take care of your food for future use.

Many vegetables as well as apples and pears can be stored effectively in their natural condition, with very little work and practically no expense. The best place to store most vegetables or fruits is in a cellar or basement where the air can be kept damp and the temperature as low as possible without actually freezing. If you don't have a suitable cellar, make use of a large closet or other areas of the house, depending upon the product to be stored. Alternatively, for simple outdoor storage, sink a box or barrel in the ground or use surface pits.

If you have a cellar without a furnace, or an outside shed, and you know that the vegetables stored there will not freeze, the organization of a storage room is rather simple. These areas can be as simple or as elaborate as you desire, with anything from a few shelves on the wall and boxes on the floor to categorized shelving and lined bins. Crates, boxes, and barrels can be used for fruits and vegetables.

A Storage Room

If you have a room in the basement that has a furnace in it, you will need to partition off the room and insulate it.

Build the room on the north or east side of the basement, if practical, and do not have heating ducts or pipes running through it.

You need at least one window for cooling and ventilating the room. Two or more windows are desirable, particularly if the room is divided for separate storage of fruits and vegetables. Shade the windows in a way that will prevent light from entering the room.

Equip the room with shelves and removable slatted flooring. These keep vegetable and fruit containers off the floor and help circulation of air. The flooring also lets you use water or wet materials (such as dampened sawdust) on the floor to raise the humidity in the room.

Store vegetables and fruits in wood crates or boxes rather than in bins.[1]

The cellar that is dug underground and contains bins filled with sand and an earth floor is a "real" root cellar. Ventilation can be provided from a window above the ground, or a pipe through the roof.

Our root cellar is a 6′ x 8′ room. You enter through a trap door and down a ladder fastened to the wall. All that rises above the ground is the square entrance and the ventilation pipe. The bins are made with large railroad ties and filled with clean, dry sand. The only objection that I have to this kind of cellar is that the spiders think the entrance and all the way down the ladder belongs to them!

Importance of Proper Temperature

Generally, the best storage conditions can be maintained in a house-cellar storage room with at least one outside window or opening for ventilation and cooling.

Apples and pears freeze at temperatures of 27° to 28° F. and most vegetables at 31° to 35° F. Freezing must be prevented,

[1]USDA, *Storing Vegetables and Fruits in Basements, Cellars, Outbuildings and Pits,* Home and Garden Bulletin No. 119 (Washington, D.C.: U.S. Government Printing Office, 1973), p. 1.

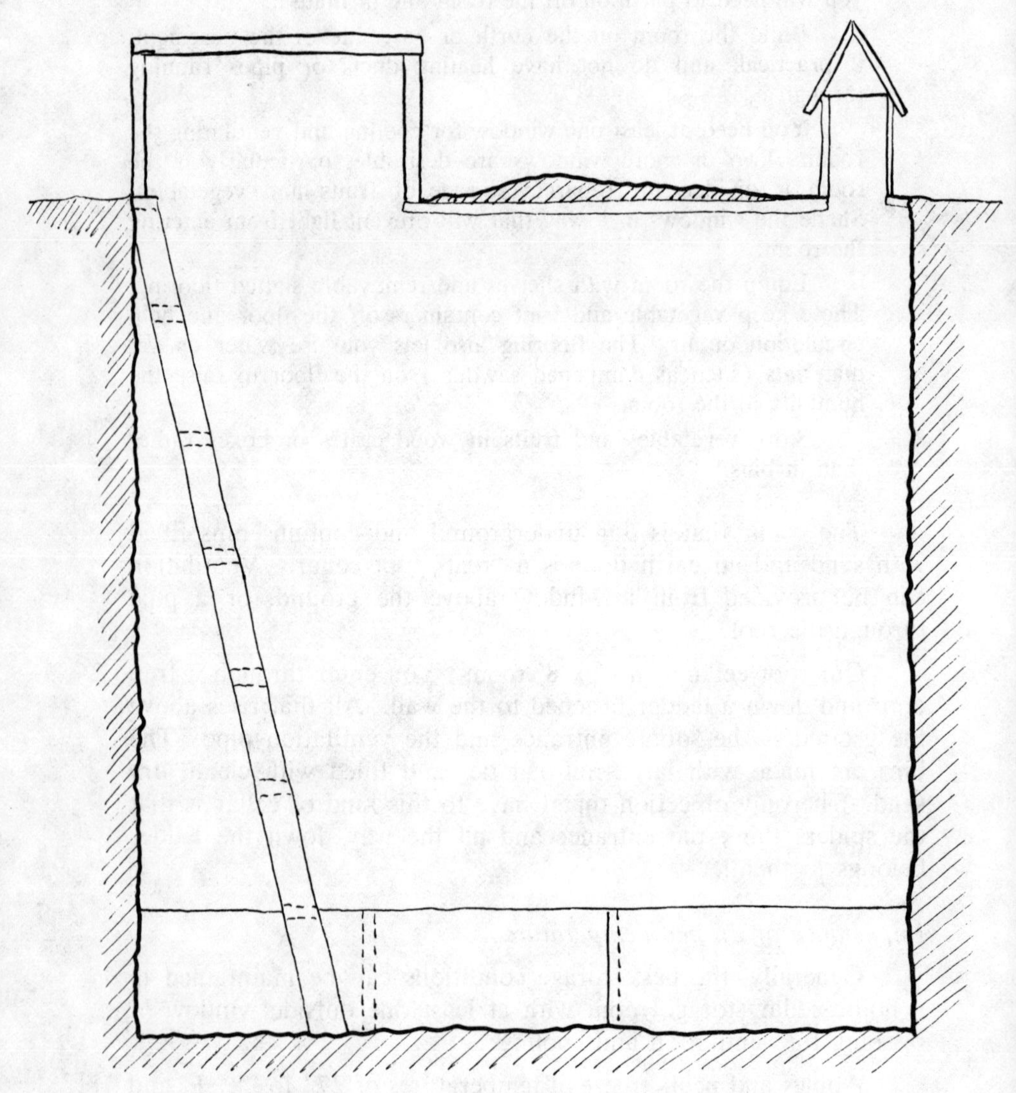

Root cellar, cross view

but the closer the temperature is to 32° F., the longer vegetables and fruits will keep.

It is almost impossible to keep the temperature this low in early fall, whereas some heat may be needed in extremely cold weather. A 100-watt bulb, hung near the floor on an extension cord, is inexpensive, safe, and usually effective for providing the necessary heat.

Need for Moist Air

The air in a root cellar must be kept moist, or fruits and vegetables will shrivel and spoil. For effective storage of most fruits and vegetables the floor should be kept damp at all times. If the room gets extremely cold, a pail of water on the floor will help prevent the produce from freezing.

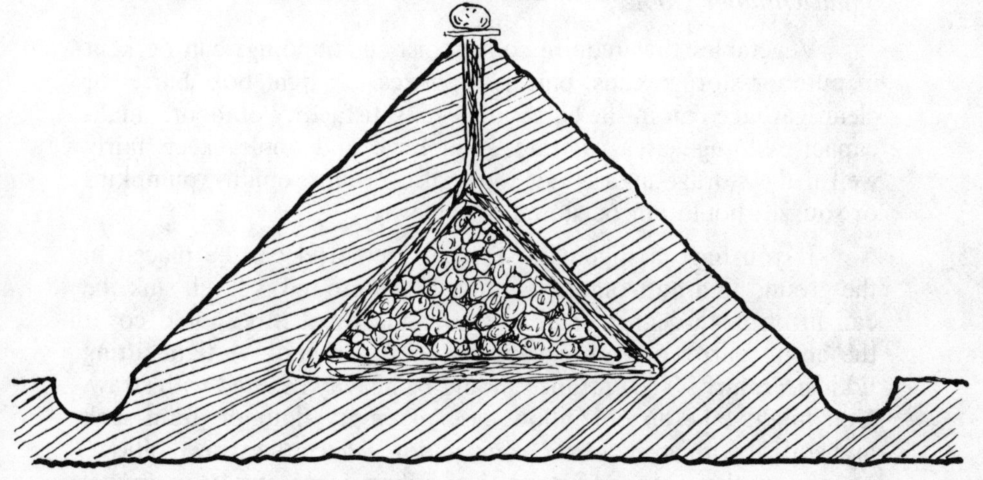

Surface pit for root vegetables

Outdoor Surface Pits

If you have access to a small yard area, you can make an "above-the-ground" root cellar. These small pits can be used to store potatoes, carrots, beets, turnips, salsify, parsnips and cabbage. Apples and pears can be stored in them too, but only for a short time.

These are shallow pits in the ground surrounded with straw or leaves and covered with earth. Note: *There must be adequate drainage*. And with such shallow outdoor pits, there is always the chance of sustaining a loss from mice.

Surface pits can be made from six to ten inches deep. Place a few inches of straw or leaves on the bottom. As the produce is put in, line the sides with straw or leaves. You can make the pile anywhere from two to four feet high. Put about a foot of straw or leaves over the vegetables, then add three to four inches of dirt.

With this type of pit or stack, there are several problems. When cold weather and snow settle in, it is sometimes hard to retrieve the produce. Also, once the pit is opened, you must take all of the vegetables out. For these reasons, it might be better to build several small pits rather than one large one. A small quantity of different vegetables could be put in each pit.

Fruits and vegetables should not be stored in the same pit.

Small Outdoor Cooler

Vegetables that require cool, moist surroundings can be kept in outdoor storage cans, barrels or boxes. A tight box, barrel or clean garbage can in the earth makes a satisfactory outdoor, small-capacity storage area. Potatoes, cabbage and apples keep fairly well if the storage area is not too moist. Beans, onions, pumpkins or squash should not be stored in this way.

If you have a slight hill, the can or barrel can be placed in the ground in a horizontal position. If the ground is level, sink the can in on a 45-degree angle. Several inches of dirt should cover the entire barrel or can, except for the open end. A tight-fitting lid is necessary. The entrance should be heavily covered with straw. You might take an old pillow case or empty flour or grain sack and stuff it with straw to act as insulation. Place the straw "pillow" on top of the produce rather than piling loose straw on top of the lid. This makes it easy to remove the produce. Simply remove the lid and the "pillow," take out whatever vegetables you wish, and replace the "pillow" and lid.

Even a wooden box with a tight-fitting cover can be buried in the ground to make a small outdoor storage area. Cover the box with plenty of straw or leaves. Alternatively, just turn a can or barrel on its side and cover it with straw and earth.

You might think these smaller versions are too small to be worth the effort. This would depend on the circumstances. For a small family, one large can full of potatoes or carrots or crunchy apples goes a long way. And potatoes purchased in hundred-pound

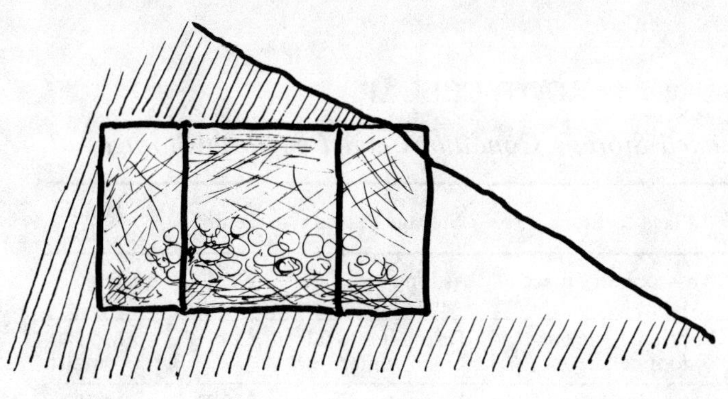

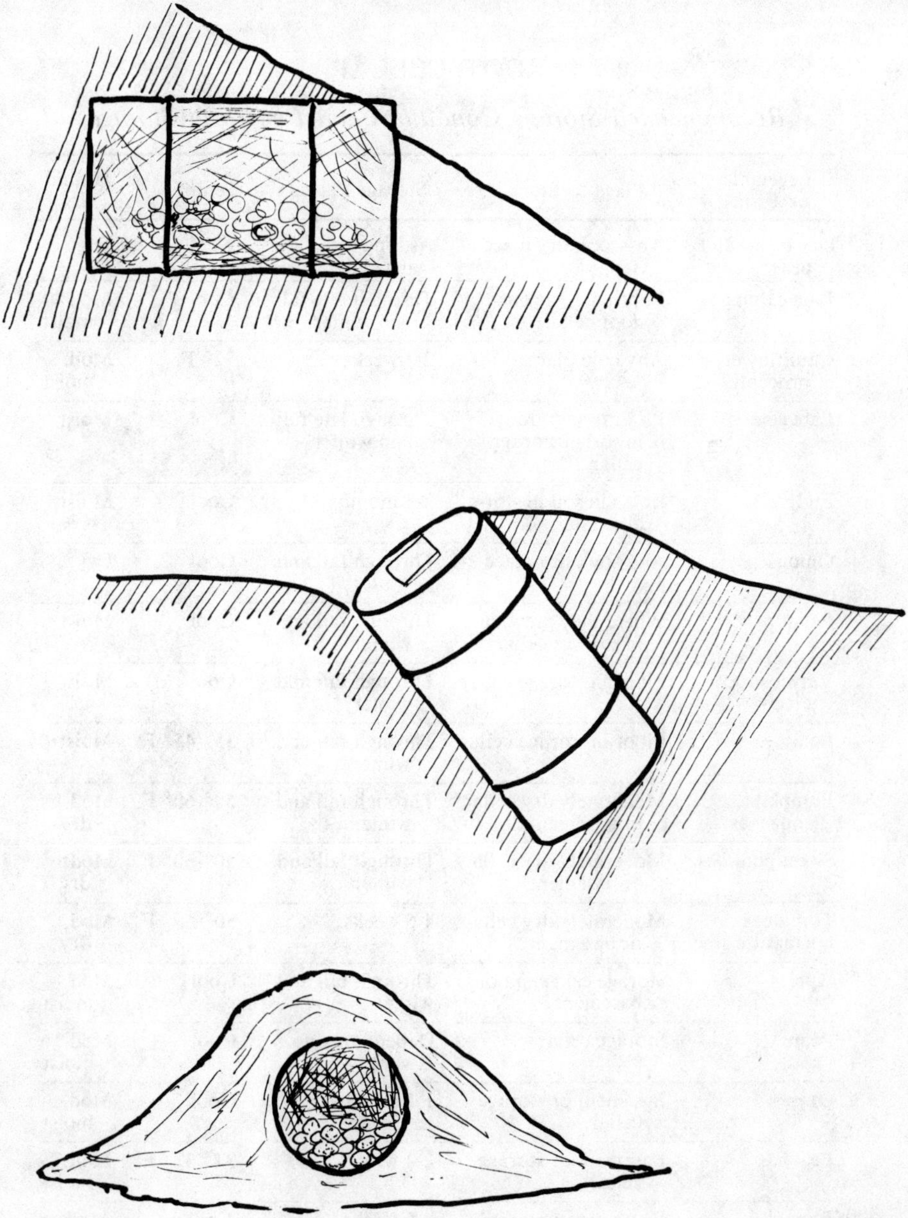

Surface storage — barrels

ROOT CELLAR

Recommended Storage Conditions and Length of Storage

Vegetable or Fruit	Place to Store	Storage Period	Temp.	Humidity
Dry beans and peas	Any cool, dry place	As long as desired	Cool	Dry
Late cabbage	Pit, trench, or outdoor cellar	Through late fall and winter	Cool	Mod. moist
Cauliflower, broccoli	Any cold place	2-3 weeks	32° F.	Mod. moist
Late celery	Pit or trench; roots in soil in storage cellar	Through late fall and winter	Cool	Moist
Endive	Roots in soil in storage cellar	2-3 months	Cool	Moist
Onions	Any cool, dry place	Through fall and winter	Cool	Dry
Parsnips	Where they grew, or in storage cellar	Through fall and winter	Cool	Moist
Various root crops	Pit or in storage cellar	Through fall and winter	Cool	Moist
Potatoes	Pit or in storage cellar	Through fall and winter	35°-45° F.	Moist
Pumpkins, squashes	Moderately dry cellar or basement	Through fall and winter	50°-60° F.	Mod. dry
Sweet potatoes	Moderately dry cellar or basement	Through fall and winter	50°-60° F.	Mod. dry
Tomatoes (mature green)	Moderately dry cellar or basement	4-6 weeks	50°-60° F.	Mod. dry
Apples	Storage cellar, pit or basement	Through fall and winter	Cool	Mod. moist
Pears	Storage cellar	Depending on variety	Cool	Mod. moist
Grapes	Basement or storage cellar	1-2 months	Cool	Mod. moist
Peaches	Basement or storage cellar	2-4 weeks	31°-32° F.	Mod. moist
Plums	Basement or storage cellar	4-6 weeks	Cool	Mod. moist

Notes:

1. Always avoid contact with free water that may condense and drip from ceilings.
2. Cool indicates a temperature of 32° to 40° F. Avoid freezing.

Based on USDA, *Storing Vegetables and Fruits in Basements, Cellars, Outbuildings and Pits.*

bags once or twice a year, rather than in five- or ten-pound lots at the local grocery store, would save money.

Things to Remember

Fruits and vegetables should be free from decay, disease, or injury and bruising that would lead to decay. Remember, as the adage says, one rotten apple can destroy the whole lot. For storage, consider only good quality produce.

Choose produce varieties that store well. These are usually the winter varieties.

Placing vegetables or fruits in storage before the start of cold weather can result in early spoilage. Root cellar storage is not long-term storage, and it takes care and watching, but it is worth the small amount of effort necessary.

Storing Vegetables

More vegetables than fruits can be stored in a root cellar. Some vegetables can be stored together, but the conditions required vary for different vegetables.

Vegetables need the temperature, humidity and ventilation controlled in order to maintain their flavor and not decay.

14
Meat Preservation

For most situations, the practicality of preserving meats at home by methods other than freezing or canning in a pressure cooker is doubtful. But in case the need should arise, information on preserving meat is presented here.

Smoking

Meat can be preserved at home by smoking. Small commercial smokers are available in most department stores and camping equipment outlets. You can make your own smoke house out of an old refrigerator or a fifty-five-gallon drum. Construct your smoke house so that you can control the ventilation and temperature. The heat should be kept at about 90° F.

Pieces of meat can either be hung on hooks or spread out on racks. Place the meat centrally in the smoker so as to insure good circulation of heat, smoke and air. Smoking time will vary for different kinds of meats.

> For kippered fish (with a high oil content) a "hot smoking" process is used with a temperature of 100°-125° F. for 10 to 12 hours, which is then increased to 150° F. for 2 to 4 hours. Smoking of lamb takes 2 days at 100 to 120 degrees. Other meats may take from one to two weeks.[1]

Brining or Curing Meat

Meat can be preserved by curing or brining, following the same basic principles as for vegetables. Many people prefer to brine meat before drying it.

---

[1]Dickey, *Passport to Survival,* p. 105.

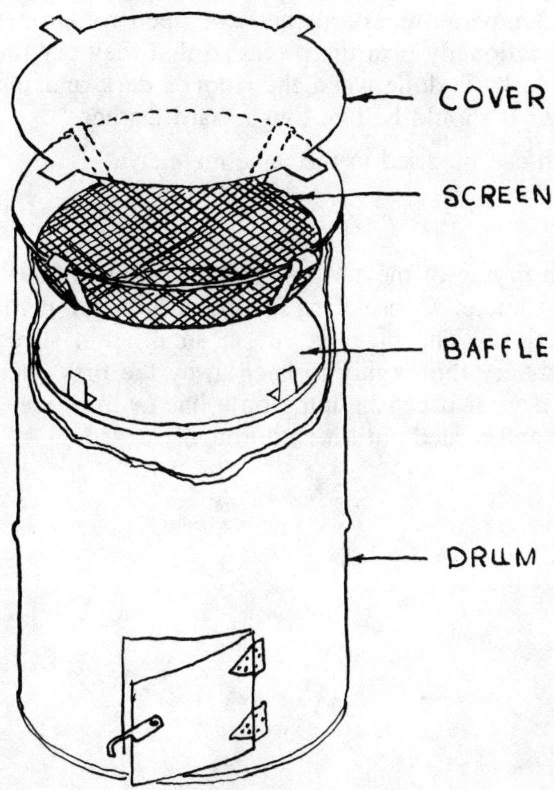

Homemade smoker

A salt compound, available commercially, makes meat curing extremely easy. This preparation can be used as a dry cure or a brine cure. This compound is also recommended for use on meat that has been frozen and is now thawing.

Oven Jerky

You can use your oven to dry meat for jerky. After removing the fat, cut the meat into slices one-fourth to a half-inch thick and one to two inches long. Cut the slices along the grain. You can pound salt, pepper and other spices into it to your taste or it can be dried plain.

Place the meat slices on racks or sheets in a thin, even layer, in a manner similar to drying fruits. Set the oven at the lowest possible temperature. Prop the door open for proper air circulation. Occasionally turn the pieces so that they dry thoroughly and evenly. Jerky is done when the color is dark and the pieces have shriveled. It should be like tough, hard leather.

Fish can be dried in the same manner.

Sun Drying

Sun drying of meat is not practical, unless you live in a very hot, dry climate. Considering the insect problem, it still may not be too practical. The meat is cut in small, thin slices, salted and peppered very thoroughly to keep away the flies, then laid out on trays to dry; or it can be hung on a line or in a tree. It should be watched and turned until it is thoroughly dry.

15
Dehydrated and Freeze-dried Foods

Two commercial processes of preserving food, dehydrating and freeze-drying, should not be overlooked. To some of you these terms may be new; others may have heard or read a little about them. But even though the terms sound strange, you have been using many of these food products for several years. They are often referred to as "convenience foods." Included are instant potatoes, dry milk, salad dressing mixes, soup mixes, cake mixes, instant onions; in fact, most instant or prepared mixes have been dehydrated.

Campers, back packers and other sportsmen have been using freeze-dried products for years.

Until recently, when more consumers became aware of and interested in dehydrated and freeze-dried foods, they were mainly used by the armed services, restaurants, schools and hospitals. They were usually packaged in institutional size (no. 10) containers, and the retail outlets handling them were scarce. But that is changing as consumer demand for these products increases.

Consider these new foods, for they are excellent used in everyday meals and have many advantages as storage items for any emergency situation.

Definitions of Processes

Technically, all drying of foods — even home drying — is done by dehydration or removal of water. But common usage has brought about some distinctions, as follows:

Dried foods or drying. This usually applies to foods which are dried at home or found in the produce section of your local supermarket. The moisture content of these products is about 20-25 percent.

Dehydrated. This is a commercial process of water removal which usually takes the moisture content down to 2-3 percent. The process is accomplished with heat.

Freeze-drying. Another commercial way of removing water, this process usually takes the moisture content to 2-3 percent. This is accomplished by a freezing process.

Low-moisture foods. This is a general term used in reference to all of the above. If it should be on a label of foods you are considering buying, read further to make sure you are getting the kind of product you wish.

Dehydrated Foods

> Dehydrated foods are usually economical. Most of them are easy to prepare and work with. Their weight and volume make them convenient to store.
>
> Much has been done to improve the quality of dehydrated foods. Some of the things that have brought about this improvement are: food that is adapted to the requirements of dehydration; new processing technology; careful application of processing procedures; less sulfuring of vegetables; improved equipment; much lower moisture content in the finished product; and packaging with inert gas.[1]

This means that the men who were in the armed services and vowed they would never eat another dehydrated egg have a surprise in store. These processed foods taste extremely good.

> Dehydration is done under a high vacuum and a very low drying temperature. This process can be done with very little loss in the overall quality of the original unprocessed product. The processing also makes it possible to remove practically all of the water from foods so that, when properly packaged, they store well.[2]

After being processed, dehydrated foods shrivel and become small, hard, and very brittle. Reconstituting them by adding water has no effect on their food value and restores the original taste and appearance. They resemble a cooked food in appearance.

---

[1]USDA, *Yearbook,* p. 424.
[2]*Ibid.*

Freeze-dried Foods

> The process of freeze-drying involves the removal of water from a product while it is frozen, by sublimation. (Which means the ice is changed to vapor without going through the liquid stage.)
>
> In the normal process, the food is first prepared in the usual manner, then frozen and then placed in a vacuum chamber; a small amount of heat is applied externally to the product in order to drive the moisture out while the food is in the frozen state.
>
> Although refrigeration is not required, the product does tend to deteriorate with long storage unless properly packaged. Low-temperature storage is advantageous.[3]

Freeze-dried foods retain their original sizes and shapes after processing.

Advantages or Disadvantages

Both types of products require a small amount of shelf space when compared by volume with canned goods.

Both products are easy to work with, and fit easily into every-day menus.

Freeze-dried foods are more expensive.

Dehydrated products shrink a great deal, therefore they allow more reconstituted weight per can.

Some foods do not dehydrate successfully, but are available freeze-dried, such as pineapple, berries, and meats.

Both dehydrated and freeze-dried products are usually dusted with sulfur dioxide to prevent changes in color. It is easily rinsed off and does not affect the flavor or food value.

Since most of the moisture and air are removed, bacteria growth is reduced and spoilage prevented.

Availability of Various Size Packages

Both dehydrated and freeze-dried products are available in various sizes and types of containers. Most companies selling these foods have basically the same size containers available.

The most popular container size is a no. 10 can, which is the same size as a one-gallon paint can. The dry weight and reconsti-

---

[3]Hughes and Bennion, *Introductory Foods,* p. 418.

tuted weight will vary with each product. This is a workable-size container for most individuals and families.

Some products are available in bulk, and bulk prices are usually more economical. You can then put the product in your own five-gallon cans or other storage containers.

The majority of these items are available only through restaurant suppliers, wholesale outlets, or home storage food suppliers. Some products, such as dry milk, dehydrated eggs, or instant potatoes, are available in the grocery store.

Proper Storage

One of the rumors about these types of foods is that if they are purchased in a sealed container, all you need to do is put them in a corner or on a shelf and forget them. This is not true! Even with an extremely low moisture content, they must be stored properly. The same rule applies as for wet-pack food: dark, dry and cool. Heat is one of the greatest deterioration factors. Keep the containers dry and off the floor to prevent moisture from condensing on the metal and causing rust.

Shelf Life

Some dehydrated or freeze-dried items have a short shelf life. Work these into daily menus and replace them. Milk and milk products, eggs and egg products are among items that will not keep for a long time. Under the proper conditions, fruits and vegetables have a shelf life of five to ten years or longer.

Once Opened, How Long Do They Last?

After you open the can, how long will the taste and food value of this food be maintained? Claims differ on this. Some assert that, although absorption of air and moisture admittedly tends to soften the food at the bottom of the can, there is no loss of food value within a reasonable period of time (up to a week or so). Others say this is not so. The taste will not usually be affected in a reasonable period of time. Naturally you will keep the opened can tightly covered with a plastic lid, because the longer the contents remain uneaten, the more moisture and air they absorb, and the more heat they are exposed to, the higher the nutritional loss.

Inert Gas

Many companies package their products with inert gas, usually nitrogen. It works the same way as home use of dry ice or carbon dioxide mentioned in chapter 3, forcing out the air and moisture and thus creating a more stable storage condition. It is not harmful to the food. Foods processed this way should still be kept in a cool, dry place.

Some Advantages

These foods are very tasty. One advantage they offer in an emergency situation is that you can continue to include fruits and vegetables in your diet. In your daily meals, they can be used the same as fresh fruits and vegetables. They are very easy to work with, but it is wise to try them now, even though you consider them as an emergency item only. You will gain experience in cooking with a variety of foods, and from it you can develop favorite recipes.

Trial and Error

When I first started using dehydrated foods, I learned by trial and error. There weren't many recipes available then. One evening we decided to have cooked cabbage, one of our favorite vegetables. I opened a no. 10 can of dehydrated cabbage flakes. This was to be our main vegetable for the meal. As I prepared the other foods, I took a handful of cabbage flakes, put them in a pan, and covered them with water. It didn't look as if there was enough, so I added another handful. A few minutes later I checked the pan and it still didn't look as if there would be enough for four of us. So I added another handful. We seemed to be eating cabbage for weeks! You can steam it, fry it, bake it. . . .

I also learned right away that dehydrated fruit is a delicious snack. Like potato chips or peanuts, once you start nibbling you just can't quit. At the time, my husband worked for one of the Southern California utility companies. There were many days that he unexpectedly would have to work overtime. I got into the habit of filling a sandwich bag with different kinds of dehydrated fruit and putting it in his lunch sack. He would put the sandwich bag in his pocket and nibble on the fruit all day. Soon many of the fellows he worked with wanted to share his snacks and were always asking, "What did you bring today?" Sometimes I would send two sandwich bags to make sure Larry had enough.

Our family is a great one for teasing, so one day I filled one small sack with dehydrated fruit and the other with a well-known brand of dry cat food. Larry, being used to my teasing, recognized it immediately. However, when one of his friends said, "What do you have today?" Larry replied, "A brand new food. Here, you can have the whole bag. See — I have another." The unknowing friend ate cat food all day, trying to figure out what the "new food" tasted like. And Larry nibbled on fruit.

Food Value

Water removal concentrates the foods and their natural sugars. The nutritional loss, for any given brand or batch, will depend upon such factors as quality and maturity of original product, length of storage before processing, and processing methods. Apparently there may be some shelf loss in food value, though that for nitrogen-pack dehydrated and freeze-dried foods is less than for any other commercial food processing. All in all these two processes produce about the best foods of their kind for emergency storage, and they are flexible enough for daily use if desired.

SOME VARIETIES AVAILABLE

(d *indicates dehydrated,* f *indicates freeze-dried*)

Fruits	Vegetables	Juice	Dairy Products	Meat
Apples (d or f)	Beans, green (d or f)	Crystals (d or f)	Butter, powdered (d)	Beef chunks (f)
Apricots (d or f)	Beets (d or f)	Lemon	Buttermilk (d)	Beef, diced (f)
Bananas	Carrots (d or f)	Orange	Cheese, powdered (d)	Beef patties (f)
flakes (d)	Celery (d or f)	Grapefruit	Cottage cheese (f)	Beef steak (f)
slices (f)	Corn (d or f)	Grape	Cream, powdered sour (d)	Chicken, chunks (f)
Dates (d)	Green peppers (d or f)		Cream, powdered sweet (d)	Chicken, diced (f)
Figs (d)	Onions (d)		Eggs (d)	
Fruit Cocktail (d or f)	Peas (d or f)		Ice Cream (f)	
Peaches (d or f)	Potatoes (d)		Milk (d)	
Pears (d or f)	Soup Blend (d)		Shortening, powdered (d)	
Pineapple (f)	Stew Mix (d)			
Plums (f)	Tomato Crystals (d)			
Prunes (d)	Yams (d)			
Strawberries (f)				

How to Use

The fruits and vegetables are very easy to use. Simply measure out the desired amount of vegetables, simmer until tender, and season. It usually takes five to fifteen minutes, depending on the hardness and thickness of the vegetable. Season these vegetables just as you would fresh ones.

If the fruits are to be cooked or stewed, season and simmer as you would fresh fruit. If the fruit is to be used as a dessert, simply cover it with water and allow it to reconstitute. Again, the time varies with the kind of fruit.

The meats must be reconstituted first and then prepared and seasoned as you do fresh meats.

The dairy products are also convenient and easy to use. Generally the cheeses are powdered or very fine granules. They can be used in sauces and other cooked dishes, or as toppings. The shortening and butter can be made into paste or creamed form. Most of you are familiar with dehydrated milk, which can be used dry, as a recipe ingredient, or reconstituted as a liquid. (See chapter 18.) The eggs can be used as an ingredient or reconstituted and used as you do fresh eggs. (See chapter 17.)

For a superb treat, try some ice cream. It's delicious! And of course it's nourishing because it is a milk product.

Shelf Space Required

Dehydrated and freeze-dried products require much less shelf space than canned or frozen foods. "Fresh frozen and canned foods contain 8-9 pounds of water for each pound of solids. Low-moisture foods contain 2-3 percent moisture per pound of solids."[4]

Three or more no. 10 cans will fit in the space required for one case of canned goods. Considering the reconstituted volume of the dehydrated or freeze-dried foods, this would be the same as having three or four cases of food occupying the same space as a case of regular canned food. (The weight and volume varies with individual fruits or vegetables and the cut and style of pack.)

Rotation

Dehydrated or freeze-dried fruits and vegetables do not have to be rotated as frequently as canned or frozen ones. Under proper storage conditions and temperatures they keep very well. However, products containing eggs, milk, or fats and oils must be rotated to insure high quality.

Economy

Several factors must be considered to determine whether these foods would be economical for you. Some of these are:

---

[4] *Low Moisture Foods* (Emeryville, California: The Vacu-Dry Co., 1973), p. 2.

Do you have access to produce at good prices to can or freeze at home? Do you have time to preserve foods at home? Do you have very little storage space? Is your preference commercially prepared food? Are there frequent employment relocations of your home? Would usage be during winter months, when most fresh produce is out of season?

The dehydrated products generally are quite economical. The freeze-dried foods tend to be more expensive. In buying these products, it is well to understand what the cost per can means in terms of servings. Six or seven dollars for *one* can of peaches seems a staggering amount until you realize that the water has been removed and you are paying for a *dry* product. Once you add water, most dehydrated food will double and sometimes triple in bulk. (Remember the cabbage?)

> For example, suppose you've allotted $5.00 for your storage budget this month to buy fruit cocktail. The supermarket has Brand X, No. 303 cans on sale at 5/$1.00. You will bring home 25 cans. When drained and measured the amount of fruit per can is approximately 1 cup. This will give you 25 cups.
>
> In contrast, one no. 10 can of dehydrated fruit galaxy weighs 44 ounces. The cost of one no. 10 can is approximately the same as the case. One cup equals 4 ounces dry weight. One can contains 11 cups dry. However, when the water is returned to the product, there are approximately 33 cups of fruit.[5]

---

[5]Barbara G. Salsbury, *Just Add Water* (Bountiful, Utah: Horizon Publishers. 1972), p. 1.

16

T.V.P. or T.S.P. – New Foods

T.V.P. — Textured Vegetable Protein. T.S.P. — Textured Soy Protein. These terms are rapidly becoming common in the food industry. Soy protein products are important. Unknown until recently to most consumers, these new products can help to supplement the diet as well as offer aid to the budget.

Soybean — the Parent

In order to understand the value of T.V.P. or T.S.P. products, let's examine their source — the soybean.

> The soybean, rich in nutrient resources, is relatively new on the American table. Before the 1900s, it was in limited production and used solely as fertilizer. In 1922, the soybean processing industry was introduced to this country, and during the following years new production techniques have allowed greater utilization of the natural protein in the soybean.
>
> First there was soy flour, then came soy sauce, soy oil, soy grits, soy lecithin extraction. The latest member of the soy family is textured vegetable protein.[1]

The soybean, a member of the legume family, is about one and one-half times higher in protein content than any other legume.

Nutritional Value

How nutritional are textured vegetable protein foods? We can find enthusiastic answers to that question.

> They are approximately equal to their animal meat or familiar counterparts. By precise engineering and control, the quantity

[1]Carma Wadley, "Stretch Your Meat With Textured Soy Protein," *Deseret News* (Salt Lake City, Thursday, December 6, 1972), p. D1.

> and quality of protein, fat, carbohydrates, calories, vitamins and minerals contained in the traditional protein foods, such as meat, poultry and fish, can be duplicated in these new foods. Textured vegetable protein is often produced free of animal fats and cholesterol.[2]
>
> Soy protein contains all eight of the amino acids considered essential for human needs. Undesirable fats can be eliminated, and calorie levels reduced for those persons on strict diets. When added to other foods, soy protein can increase the total protein content and improve its quality.[3]

But warning voices are being raised. Some suggest that, apart from their objection to such things as the chemical additives used in the processing, there may be problems related to the proper assimilation of the nutrients. At time of writing it is clear that the advantages and disadvantages of this relatively new product are far from being fully determined.

The answer seems to be that T.V.P. probably is valuable in moderate amounts, certainly for emergency storage.

How Is It Made?

Several processes may be used for making T.V.P. In one, the protein is extracted from the soybean and spun into fibers, much like nylon is spun. These fibers can be fashioned into various textures, and are often made to have the texture of various meats.

In another process, these products are made from soy flour, colored and flavored, then extruded through special equipment to resemble a variety of products, from beef granules to nut meats. The most popular product of this type is used to supplement ground meats.

There are several new frozen varieties, now or soon to be on the market, that look, taste, and feel like meat.

Soy protein products are not intended to replace meat. A granule could never replace a T-bone steak! Keep in mind that these are new foods — not "imitation" products.

---

[2]*Everything You Want to Know About Bontrae, Textured Vegetable Protein* (Minneapolis: General Mills, 1973), p. 2.

[3]National Soybean Processors Association, *The Story of Soy Protein* (Washington, D.C.: Food Protein Council, 1973), p. 3.

Advantages

Variety, economy, good nutritional value, and convenience are a few of the virtues of T.V.P. Economy is a key factor in their popularity. Says one processor: "As a meat analog, textured protein offers the processor, as well as the consumer, important new economies — one-third to two-thirds the cost of cooked meat."[4]

These foods can be used in any diet. Their versatility is shown by the fact that they can be used to extend meat in your favorite recipe, or they can be used alone in the same recipe. Since T.V.P. contains no fat, it will not shrink. In many cases it doubles and triples in bulk as it absorbs juices and liquids. Use T.V.P. as you would ground beef, chopped ham or diced chicken. The unflavored kind will absorb the flavor of the sauces and spices that it is cooked in, or it can be flavored with bouillon, soups, or juices.

T.V.P. products are convenient to use and are easily worked into most recipes. Thus, though one of the ingredients is new to the cook, the taste will be familiar to the family.

Storage

The dehydrated or dry T.V.P. will store eighteen to twenty-four months, if properly cared for. The frozen style will keep six to nine months in a deep freeze. Dry T.V.P. can be purchased in bulk in heavy paper sacks lined with plastic, but this sack is not sufficient protection for storage. If kept in the sack, T.V.P. will absorb moisture from the air, which will cause it to become stale rapidly. Proper storage for T.V.P. is an airtight container in a cool, dry place.

How Much to Buy

Be aware that T.V.P. is purchased differently from meat. Ground meat, for instance, is bought by the pound. With T.V.P., however, the yield is much greater.

One cup of dry burger granules would equal the *bulk* of one pound of ground meat. A pound of the dry T.V.P. granules will yield approximately the same *bulk* as three pounds of ground meat. The yield naturally varies with the type of T.V.P.; that is, granule, dice or chunk.

---

[4]*Textured Protein Systems, New Food For Thought* (Minneapolis: General Mills, 1973).

Suggestions for Cooking with T.V.P.

Any new product is "different" until *you* become familiar with it. Attitude makes a big difference. Be patient, and don't be afraid to experiment. If you get a sloppy joe mixture instead of meat loaf the first time you try using it (and you were using a meat loaf recipe), try again!

Once you have become acquainted with T.V.P., it likely will become a staple item in your cupboard and a regular on your menus. It cuts down preparation time, and if it is prepared properly it is difficult to tell that a meat substitute has been used.

There are some of the ways these products can be used:

As a topping
In ground meat
In sandwiches, chowders, soups, salads
For pot pies
Creamed

APPROXIMATE EQUIVALENT CHART

Dry Weight		Wet Weight (water added)		Fresh Weight
1-1½ cups TVP burger granules	=	2-3 cups granules	=	Bulk of 1 lb. ground beef
1 cup T.V.P. ham chunks	=	2 cups "ham"	=	Bulk of 2 cups cooked cut-up ham
½ cup T.V.P. bacon bits	=	Do not add water to reconstitute	=	Bulk of 3-5 slices of bacon cooked crisp and crumbled.
1 cup T.V.P. chicken chunks	=	2 cups "chicken"	=	Bulk of 2 cups of cooked cut-up chicken

Basic Directions

Barely cover the T.V.P. with warm to hot water. (Add bouillon for a much better flavor). Allow to stand 10-15 minutes to absorb moisture and flavor. Then use as meat in recipes.

The burger granules tend to be dry when using them in recipes such as meat loaf. Make sure that you use enough binding liquids to hold them together. (Bindings such as cream soups, catsup, eggs, etc.)

T.V.P. bacon bits should not be soaked, but are added directly to the recipe being prepared.

If the T.V.P. product is being added to recipes such as soups, chowders, or casseroles which have ample liquid in them, add the T.V.P. directly into the mixture in dry form.

Be sure to use seasonings and spices. Compared to "real meat," these soybean products tend to be bland.[5]

You can see that T.V.P. products make a good storage item for the pantry. They will help to make meals delicious if hard times decide to visit you. Or perhaps they would stretch the food budget now to help you make ends meet.

[5]Salsbury, *Tasty Imitations,* p. 22.

17
Eggs

The egg has been thought of as a symbol of good fortune by the peoples of many lands since ancient times. To many primitive people, the egg signified the return of life following the long winter season. Many are the traditions and legends of the egg. No doubt the use of eggs in celebrating Easter by people in Christian lands is a tradition handed down from earlier beliefs.

In America today, eggs are considered an important food. Fresh eggs are so much a part of our daily lives that we often forget how valuable they are. "In addition to their uses, eggs are universally available, 100 percent natural and attractively packaged in convenient, pre-measured amounts."[1]

Eggs are an excellent and inexpensive source of protein. They also furnish significant amounts of iron, vitamin A and riboflavin (vitamin B_2).

Uses for Eggs

Eggs are the basis of many dishes, whether served alone or in other foods. Here are just a few of the ways they are used: In appetizers, salads, salad dressings, breads, sandwiches; in combination with vegetables and cereals; as an ingredient in desserts, cakes, pastries, cookies, and custards. For serving alone, they are boiled, poached, fried, scrambled, baked, deviled, or made into omelets.

Do you think you can cook very much without an egg or two?

---

[1]"Take a Fresh Look at Eggs," *Good Food* (Radnor, Pennsylvania: Triangle Publishing Co., 1974), p. 47.

How Eggs Function

Understanding the functions of eggs in a recipe will help you to see why they are important.

Among the many roles eggs have in food preparation are these:

— Thickening: in custards, puddings, fillings and sauces. Two egg yolks or two egg whites have the same thickening power as one whole egg.
— Leavening: for cakes, souffles, and omelets.
— Blending or binding: binds ingredients, as in meat loaf, or holds crumb coatings on foods.
— Emulsifier: blends and stabilizes two liquids, such as oil and vinegar in mayonnaise.

Eggs should be considered top priority for the pantry. Several ways of preserving them will be discussed here.

Frozen Eggs

Eggs can be frozen whole, or whites and yolks can be frozen separately. Only clean, fresh eggs should be used for freezing. Do not use chipped or cracked eggs. Be sure that strict sanitary conditions exist in all preparations for freezing eggs. Wash your hands thoroughly and make sure that all utensils to be used are clean.

Egg yolks are changed by freezing and tend to become stiff and gummy when thawed. To prevent this, a small amount of salt, sugar or corn syrup may be added before freezing.

Break eggs one at a time into a cup before putting them into a clean bowl. Discard any eggs that have an odor or dark spots.

For *whole eggs,* gently stir the eggs with a fork to mix them together, but do not beat them. For each cup of liquid whole eggs add:

1 Tbsp. light corn syrup or sugar
or
½ tsp. salt

How you are going to use the eggs will determine whether you add salt or sweetening. Sweetened eggs are best used in baking or desserts such as cakes, custards or pies. The eggs with salt may be used for scrambled eggs, main dishes, or foods made without sweetening.

Egg whites can be frozen without adding anything. There is no need to stir them. Remove any yolk with a spoon, since even a small amount will prevent the whites from beating well. "One drop of yolk will reduce the volume of whipped egg to less than one-third of that otherwise obtained."[2]

When packaging eggs in a freezer container, remember to leave headroom for expansion.

One idea of packaging is suggested by the Utah State University Extension Service. It may be convenient to freeze egg whites, or whole individual eggs, in small plastic ice cube containers which will hold one egg or egg white. When they are frozen hard they can be put into a plastic bag and used as desired.

Be sure to label each container, indicating how many eggs are in it and whether they contain salt or sweetening. The containers must be moisture- and vaporproof to insure good quality.

Eggs spoil easily when they become warm, so it is best to thaw frozen eggs in the refrigerator. If egg whites are to be whipped, allowing them to reach room temperature will result in greater volume. Thaw only the amount to be used at one time, and do not refreeze eggs. Frozen eggs may be kept safely for up to twelve months.

Eggs that have been frozen should be used only in foods that require thorough cooking.

Equivalents

3 Tbsp. of frozen whole eggs equals one egg

1-1½ Tbsp. of frozen yolk equals one fresh yolk

2 Tbsp. frozen egg white equals one fresh egg white

5 whole eggs
or
12 yolks
or
8 whites } equals 1 Cup

Water Glass

Water glass is an old-fashioned way of preserving eggs at home. As a child, I can remember going down into my grand-

[2]Justin, Rust, and Vail, *Foods,* p. 212.

mother's cool and mysterious basement. She always had a beige crock full of eggs in water glass. Both the egg crock and a water bucket with a metal dipper are part of fond memories of an Ohio farm.

Water glass is sodium silicate, a clear, syrupy liquid which closes the pores of the eggshell. Sodium silicate is relatively inexpensive and can be purchased in a drugstore or hardware store. When diluted with nine to ten parts of water and mixed thoroughly, it is used as a liquid in which to store fresh eggs. Use clean, fresh, unwashed eggs. Pack the eggs (in the shell) in a stone or glass jar, or a wooden keg, keeping them well covered with the solution at all times. A tight-fitting lid will prevent evaporation of the liquid. Store in a cool dark place.

Though eggs stored in this manner tend to lose quality with long storage, they will keep nine to twelve months. Eggs stored in water glass eventually will develop a slightly stale taste when cooked. The yolks are apt to break if eggs are used for frying. After several months in the solution, eggs may develop a "sulphur" odor if they are hard-boiled. I noticed with one "test" jar of eggs, stored over ten months, that the whites became runny. The eggs still smelled and tasted like eggs, but weren't suitable for meringue or in other recipes where high volume is desirable.

This method is not for long-term storage. Eggs in water glass must be watched and cared for, not set on a shelf and forgotten. This method of preservation is probably most successful if eggs are purchased when the supply is ample and the price reasonable, the object being to have enough eggs to "tide you over." In ordinary use, water glass storage would supplement limited refrigerator space.

As part of a lab experiment for a home storage class, I once put several dozen eggs in water glass in a gallon jar. When I brought the jar out of the closet to take it to class and demonstrate how well the eggs were keeping, I discovered a cracked egg floating, with some of the egg protruding from the shell. Conjuring up the drastic odors that might assail me, I could hardly muster enough courage to open the lid and remove the offender. But I couldn't take a crock of eggs for "show and tell" with a broken one floating in plain view. Finally, I took the jar outside and opened it, ready to run if the odor was too bad. Surprise! There was no odor. The part of the egg that had come through the shell had solidified. It looked and felt like crystallized salt. The egg cracking and "leaking" had affected neither the solution nor the other eggs.

Dried Eggs

Dried whole eggs, dried egg white, dried egg yolk and dried egg mix are all available commercially. About 90 percent of the water is removed during the drying period, so that one pound of dried egg is equal to thirty-six to forty average-size eggs.

When you are considering purchasing a dried egg product, read the label carefully. Dried egg mix is a blend of whole dried eggs, nonfat dry milk, corn oil, color, salt; in some brands lecithin and/or whey are added. The egg mix costs considerably less than the whole egg solids, but your choice might depend on the intended use of the product. Egg mix can be used as an ingredient in baking or for scrambled eggs. Dried whole egg solids can be used as a replacement for liquid whole eggs in most recipes.

Cooking with Dried Eggs

For most baking recipes, the dried egg is sifted with the dry ingredients. To replace the water removed from the egg in drying, add water to the liquid in the recipe. In other recipes, usually those calling for a beaten egg, the dried egg is blended with water first, then used as you would fresh eggs. Where dried egg is used for binding, such as in meat loaf or patties, reconstitute the egg first, then use as you would fresh eggs.

In a recipe calling for beaten eggs, mix a small amount of warm water with the egg to make a smooth paste, then add the remaining water. The egg blends much easier when it is mixed this way.

To measure dried eggs effectively, put whole egg solids in standard measuring cups or spoons. *Pack firmly and level.*

Once you realize how easy and economical dried eggs are to use, they will probably find a permanent place in your working cupboard as well as a shelf place in your storage.

Equivalents

2½ Tbsp. dried egg plus 2½ Tbsp. water = 1 shell egg

1 cup dried egg plus 1¼ cups water = about 8 whole eggs

Availability

Dried eggs are available from local wholesalers, bakery supply houses, home storage food outlets, and some grocery stores. Available at most of these outlets are no. 2½ cans weighing twelve ounces, and no. 10 cans weighing three pounds. Although the price of dried eggs fluctuates along with other food products, the price generally compares with the cost of fresh eggs. It may be advantageous to use dried eggs during the winter when the cost of fresh eggs soars. They do save storage space, taking only a fraction of the space needed for shell eggs.

Storage

Dried eggs can be stored for one to two years in sealed cans, if they are kept dry and cool. The ideal storage temperature is under 70° F.

Once a can is opened for use, keep it tightly covered in a cool, dry place. An opened container of dried eggs should be used as soon as practical. Exposure to moisture causes dried eggs to become lumpy and develop a strong flavor. If stored in the refrigerator, cover the container tightly or the eggs will absorb moisture and odors.

If not kept cool, dried eggs will lose their solubility and thickening power, and the flavor could become objectionable.

In summary, dried eggs are an excellent way to store a high protein food — just in case.

18
Milk

Milk has often been called the perfect food, probably because it serves as food for the babies of all mammals. Milk, a good source of high-quality protein, calcium, riboflavin and vitamins A, D, E, and K, seems to be the one food for which there is no adequate substitute. All of the vitamins known to be essential in human nutrition are present in milk, and it is a rich source of minerals. In short, milk is an important food in our diet.

The necessary amounts of milk suggested by the U. S. Department of Agriculture for one day are three to four cups for children; four or more cups for teenagers; and two or more cups for adults.

Dry Milk

With so much importance placed on milk, we would indeed be wise to have an adequate amount on hand. Since it is not practical for each of us to maintain a cow to supply the milk we need, we must have another source. From a storage viewpoint, the best possible source is dry milk. New methods of processing have made it possible to preserve maximum fresh milk flavor and to have a product that is easy to use.

> *Regular nonfat dry milk* is usually made from fresh pasteurized skim milk by removing about two-thirds of the water under vacuum and then spraying a fine mist of concentrated milk into a chamber of hot filtered air where it dries almost instantly and drops to the floor as powder. This process produces a fine powder of very low moisture content, about 3 percent.[1]

To form *instant dry milk,* "milk powder may be treated with an additional instantizing process so that it will disperse in water

[1]Hughes and Bennion, *Introductory Foods,* p. 206.

more readily. The substance is remoistened and redried to form larger agglomerated particles than regular nonfat dry milk."[2] This process greatly increases the ease of dissolving in water. Dried whole milk is made from fresh whole milk with the water removed by the same process as is used for nonfat dry milk. "Dried whole milk contains not less than 26 percent milk fat and not more than 5 percent moisture."[3] Because of the fat content, dried whole milk does not keep as well as nonfat dry milk.

Another dried dairy product is dried buttermilk. This is used commercially in flour mixes.

Food Value

Remember that the fat has been removed from the dry milk, so it cannot be fairly compared to fresh whole milk. There may be other differences too, for while we would expect reconstituted nonfat dry milk to have the same nutritional value as whole milk less that of the fat content, it is claimed that some experiments indicate that it is not quite that simple.

Clearly the full story on the food value of dry milk has yet to emerge. In the meantime we can only treat this item as a highly important one for emergency storage.

Availability

Dry milk is widely available in grocery stores, wholesale outlets, bakery supply houses, and home storage outlets. It is one of the most economical forms of milk you can buy, with regular nonfat dry milk usually less expensive than the instant. Dry milk can be purchased in package sizes up to a hundred pounds.

Generally the type of dry milk available in the local grocery store is the instant kind. Dry whole milk is marketed only on a small scale, chiefly for infant feeding. Dried buttermilk, available at bakery supply houses, is also available in the grocery store in two- to three-pound boxes.

How to Use

Dry milk is convenient to use. It can be reconstituted with water and used as liquid milk. In recipes for breads, cakes, and other baked goods, the dry milk can be mixed with the other dry

---

[2]*Ibid.*
[3]*Ibid.*

ingredients. Water is then used for the required amount of liquid. "A good rule to follow in deciding how much dry milk to substitute is to use three to four tablespoons of dry milk and one cup of water as the liquid to replace each cup of milk called for in the recipe."[4]

However, when making bread, if you use the instant powdered milk you will have better bread if you liquefy the milk and then scald it as you do the fresh fluid milk.

For gravy, white sauce and cream soup, stir the powdered milk into the flour or blend in the dry milk after heating the fat and flour together. For puddings, mix the dry milk with the cornstarch and sugar.

Adding dry milk to a recipe is a good way to increase the nutritional value of the food we prepare. Dry milk can be added to meat loaf, casseroles, soups, etc.

Nonfat dry milk can be made into a whipped topping. A bowlful of home-canned peaches with a fluffy topping may be just the thing to cheer up an otherwise dreary day. Use a bowl large enough to hold the amount of topping desired. The bowl and beaters should be well chilled. Using an equal amount of dry milk and water, place the water in the bowl, sprinkle the dry milk over it, and beat at a slow speed until the milk is thoroughly blended, then beat at a high speed until the mixture is stiff. A small amount of sugar and flavoring can be added when the milk is beaten stiff. This topping does not hold up for a long period of time, but it is a tasty and nutritious menu brightener.

Mixing the Milk

Equivalents:

Nonfat dry milk — ¾ to 1 cup dry milk and 1 quart water = 1 quart liquid milk.

(1 pound usually makes 5 quarts of liquid milk)

Instant — 1⅓ cups dry milk and 1 quart water = 1 quart liquid milk.

When mixing instant milk, sprinkle the milk solids on top of the water, then beat or mix thoroughly, or put the mixture in a tightly covered jar and shake it vigorously.

---

[4]Marion Bennion, "Cooking With Dry Milk," *Relief Society Magazine*, February 1956, p. 402.

Regular nonfat milk is a little more difficult to mix, unless you know this easy way. Using a small bowl or cup, fill this with very warm (not hot) water; add the dry milk to it a little at a time and stir to form a smooth paste. You may then blend the paste into the remaining amount of the very warm water. This method will give you lump-free, well-blended milk. Of course you need to think ahead if the milk is to be used for drinking purposes, since it should be chilled thoroughly after mixing.

A blender is a great help in mixing milk, but it is not a necessity. An electric mixer or hand beater will serve the purpose just as well. With a little patience, even a wooden or other large spoon and a large pitcher works well.

For a better drinking flavor, add a little more dry milk than is recommended. This insures not only a better taste but more nutrition. In order to take away any "chalky" taste dry milk might have, try adding a teaspoon of sugar per half-gallon of reconstituted milk.

Storage Facts

Since it does not require refrigeration and takes up very little shelf space, dry milk is convenient to store. However, the heavy sacks that the milk is packed in are not good storage containers. It requires a tightly covered container, which will protect it from the air. Because dry milk absorbs moisture easily, it can become off-flavored and caked when exposed to the air.

Dry milk that is stored in airtight containers at 40° F. will keep for as long as two years. At 70° F. it will keep for twelve months, while it will keep only three months if stored at 90° F. Obviously, heat causes milk to deteriorate very rapidly. Store it in as cool a place as possible, away from furnaces and water heaters.

If dry milk becomes very hard during storage, just break it up into chunks, reconstitute it and use it for cooking. If stored for an extended period of time, dry milk tends to go stale and lose flavor. In that case, use it for cooking.

Try to develop a habit of using dry milk regularly. This not only will help to keep your supply in constant rotation but will also develop your knowledge for using it.

Dry whole milk does not store for an extended length of time. Once the container has been opened, it should be tightly covered and stored in the refrigerator.

Canned Milk

Another way to have a supply of milk on hand at all times is to store canned milk. This is one storage item that needs continual attention and proper care.

Evaporated milk is prepared from whole milk by the evaporation of water. "Most of the evaporated milks on the market are fortified with 400 U.S.P. units of vitamin D concentrates per quart."[5] By mixing equal parts of evaporated milk and water, you have a product that can be used as you would fresh liquid milk.

Evaporated skim milk, one of the newer milk products, can be diluted with an equal amount of water and used like fresh skim milk.

Sweetened condensed milk is used almost entirely in making candy, cookies and desserts.

Cans of evaporated milk (including sweetened condensed milk) should be turned every few weeks during storage because the solids tend to settle and separate.

The best storage temperature for evaporated milk is 40° F. At this temperature it can be stored for a year. It should not be frozen. Evaporated milk develops a brownish color as the storage time lengthens.

If evaporated milk has separated, shake the can very hard before opening, and it will become smooth again. Sometimes canned milk is watery and lumpy when it is poured out. This does not necessarily mean that it has spoiled.

Canned milk is easy to use in recipes or as a liquid milk. It tends to have a heavier, creamier taste than regular milk.

Making Topping

Evaporated milk can be made into a whipped topping. Chill undiluted evaporated milk to about 32° F., until fine ice crystals form. The bowl and beaters should be chilled well. Add two tablespoons of lemon juice or vinegar for each cup of milk, for greater stiffness. This evaporated milk topping will keep up to an hour if it is kept chilled.

---

[5]Hughes and Bennion, *Introductory Foods,* p. 206.

19
Cheese

Cheese is one of the oldest milk products known to man. Where it was first used is not known, but many legends of its origin have been handed down through the centuries.

Not many decades ago cheese making was largely a home process. Today most cheeses are made commercially.

There are many varieties of cheese. The variations depend on the type of milk used, the method of making curds, the method of ripening, and the seasonings used.

Cheese resembles milk in food value. However, since a portion of the water is removed in the manufacturing process, it is a more highly concentrated food. A pound of cheese contains the protein and fat of approximately a gallon of milk. Pound for pound, cheese is higher in protein and fat content when compared with other common protein-rich foods, yet is often less expensive. Cheese that contains milk fat is also a very good source of vitamin A. It is also high in calcium and in vitamin B_2 (riboflavin). Since cheese is high in protein and fat, it can be used in place of meat and eggs in the diet.

Whether a cheese is designated as "soft" or "hard" is determined to a large extent by the amount of water it contains. A soft cheese can retain as much as 50-75 percent of water. To form a hard cheese, the water may be extracted by the use of heat, pressure or a combination of the two. The water content of hard cheese is usually 25-35 percent.

How to Store Cheese

Most cheese stores well. How long it will keep depends on the kind of cheese and how it is wrapped. All cheeses should be stored tightly wrapped. Cottage cheese should be used within two or three days.

Soft natural cheeses will store from one to two weeks in the refrigerator. Processed cheeses will store from three to six months in the refrigerator. Cheese spreads and cheese foods keep well without refrigeration until the container is opened; then the unused portion should be kept refrigerated.

Properly wrapped hard cheeses will keep for several years at refrigerator temperatures. Since the elimination of air is vital, they should be wrapped tightly to protect them from exposure to air and drying out.

Blocks or horns of hard cheese that are dipped in paraffin and stored in a cool (below 70° F.) place will keep for several months. "Five-pound bricks of cheese may be stored by wrapping in vinegar cloths. They will not mold and will not need to be rewrapped for six to eight months."[1]

Any mold that forms on the surface of hard cheese can be trimmed off before using. If cheese has dried and become hard, it may be grated, stored in a tightly covered container, and used in cooking.

Frozen Cheese

Cheese can be frozen, but texture and appearance may change. Sometimes, depending on the cheese, flavor changes are noticeable.

Soft cheese can be frozen for up to one year. Hard cheeses will keep for twelve to eighteen months when frozen. Hard cheese becomes crumbly after it has been frozen.

Cottage cheese and processed cheeses should not be frozen. One handy way to freeze hard cheese is to grate it first and package it in amounts that will be required for a particular use (macaroni and cheese, souffle, etc.).

Commercially Canned Cheese

Canned cheese is available at some food storage outlets. Usually it comes in a no. 10 can weighing approximately three pounds. This is a grated "cheddar type" cheese for use in cooking. If stored in a very cool place, it would have a shelf life of twelve to eighteen months. Compared to bulk cheese it is rather expensive, but it would be another way to keep cheese on hand.

---

[1]Bob Zabriskie, *Family Storage Plan* (Salt Lake City: Bookcraft, 1966), p. 48.

With the costs of food on the rise, it might be a good thing to know how to make cottage cheese at home. Here are two ways:

Recipe for Homemade Cottage Cheese No. 1

Allow milk to stand in a warm kitchen until thick and clabbered. Cut into small cubes with a knife. Allow it to stand undisturbed for several minutes or until the whey has been fairly well forced out.

Heat, stirring gently, to about 95°-98° F. Let it stand at this temperature until it is fairly firm to the touch, then pour through several layers of cheesecloth or other clean cloth and allow it to drain for an hour or two. (A cloth bag works better than just a cloth, but a cloth lining in a strainer or sieve will work.)

Add salt and/or pepper to taste. Sweet or sour cream may be added. Mix gently with a fork.

Homemade cottage cheese should be used within one or two days.

Homemade Cottage Cheese No. 2

(This is an easy recipe. Raw milk is good for this one, but regular milk will do.)

Heat a quart of milk in a heavy utensil or double boiler. When the milk is warm, add one tablespoon of lemon juice, stirring gently and keeping it over low heat. When the milk curdles, remove the pan from the heat.

Pour into a cloth bag or through several layers of cheesecloth, and allow it to drain for several hours. Salt and pepper to taste. You can mash the curd with a fork if you prefer a smaller more uniform curd.

Cottage cheese is good also with one or more of the following mixed in: chopped pimento, chopped pepper, chives, chopped olives, drained crushed pineapple.

I have found it more difficult to make cottage cheese with raw milk than with "store bought" milk. It takes more care and patience.

Homemade Hard Cheese

Hard cheese is more complicated to make at home than cottage cheese. A curd is formed, the same way as for cottage cheese, then it is cooked, firmed, drained, salted and pressed. Pressing is the step that usually prevents the processing of this type of cheese at home. Proper equipment is needed in order to obtain a good product. Home cheesery kits that contain all of the necessary equipment and instructions are available. These kits usually contain a thermometer, rennet tablets, press, crock, and coloring tablets.

Cooking with Cheese

As cheese is heated, it softens and finally melts. If it is overcooked, it becomes tough and stringy. Dry heat causes cheese to smoke and give off a disagreeable odor. Cheese also scorches easily, so use a heavy pan and low temperatures for success in cooking with cheese.

Use a mild cheese with mild-flavored meats, fish or vegetables. The stronger cheeses go better with stronger flavors. Cheeses blend well with most herbs and spices.

Having a wide range of uses plus extremely high food value, cheese should be kept in good supply. Its excellent keeping qualities make this an easy thing to do. If you become familiar with the many uses of cheese, the variety of flavors and textures can add a good deal of strength to your menus.

Suggested Uses for Cheese

Soups	Spreads
Sauces	Breads
Casseroles	Pancakes
Main dishes	Salads
Sandwiches	Fondue
Toppings	Souffle
Side dish	Nibbles

Combined with pasta, potatoes, meat or vegetables

20

Fats and Oils

Cream, butter and fat have been part of man's diet since history has been recorded. The phrase "living off the fat of the land" indicates the high regard in which they were held. Pioneers in every civilization have found resourceful ways of conserving and using fats. They were even used for bartering in international trade. Today fats and oils are commonly used in baking, cooking and frying.

Purpose in Cooking

> The primary purpose of shortening in pastry, biscuits, muffins, and cakes is to make a tender product. This effect may arise because a thin film of fat forms around the particles of other ingredients and the gluten cannot form.[1]
>
> Shortening makes the gluten meshwork of the dough more elastic so that gas can expand freely and easily. This helps increase the volume. Shortening improves the flavor, makes bread more tender and helps give it a soft, velvety crumb. It also makes it brown better and stay fresh longer. Any kind of fresh, sweet fats or cooking oils may be used in breads or rolls.[2]
>
> We use fats and oils for frying and cooking foods; in mayonnaise and salad dressings; and to give richness and flavor to such cooked foods as vegetables and meats.[3]

Storage and Care

It is evident that fats and oils are extremely important in food preparation. Therefore it is necessary to know how to care for

---

[1]USDA, *Yearbook of Agriculture,* p. 503.

[2]Elna Miller, *Yeast Breads Both Plain and Fancy* (Logan, Utah: Utah State University Extension Service, 1971), p. 7.

[3]USDA, *Yearbook of Agriculture,* p. 503.

them properly in order to maintain a good supply of the kinds you prefer to use.

Most fats and oils need protection from air, light and heat. They can eventually go rancid even if the container is kept tightly closed, unless they have been hydrogenated or stabilized.

Butter, Margarine, Fat Drippings

Butter, margarine and fat drippings tend to become rancid more quickly than other fats and oils because they contain more moisture. They must be covered or wrapped tightly and kept refrigerated. Exposure to heat and light hastens rancidity.

Butter and margarine can be frozen, but tend to be crumbly rather than creamy when thawed.

In a cold storage room where the temperature remains 60° F. or below, a good-quality margarine can be stored for several months.

Cooking and Salad Oils

Oils need to be kept in a dark, cool place. Exposure to light and heat will cause them to go rancid rapidly. Properly cared for, unopened oil will store for six to twelve months. Some cooking and salad oils become cloudy and solidify when refrigerated, but this is not harmful. When returned to room temperature they again become clear and liquid.

Mayonnaise and Other Salad Dressings

Commercial mayonnaise and salad dressings store well if kept in dark and cool surroundings. They will keep for twelve to eighteen months under proper conditions. After opening, however, they must be refrigerated. Refrigerate all homemade salad dressings.

Hydrogenated Shortenings and Lard

Hydrogenated shortening and lard refers to the firm, creamy style (solid) shortening. It has been "stabilized by hydrogenation (which means adding hydrogen to unsaturated fats) or antioxidants."[4] These shortenings and lard can be stored at room temperature. If stored in a cool and dark place they will keep for several years. There are suggestions however that hydrogenated products inhibit the body's assimilation of essential fatty acids.

---

[4]USDA, *Storing Perishable Foods in the Home,* Home and Garden Bulletin No. 78 (Washington, D.C.: U.S. Government Printing Office, 1973), p. 6.

Peanut Butter

Peanut butter in a glass container, unopened, will keep for six to nine months if kept dark and cool. Peanut butter is also available in a no. 10 can, which holds several pounds. The cream style will keep for eighteen to twenty-four months under proper conditions. The crunchy style with peanuts mixed throughout does not store as well, since the peanuts tend to cause rancidity more quickly. It will keep for nine to twelve months if stored properly.

21
Sweeteners

Without sweetness of some sort, our diets would be drab indeed. I do not wish to get involved in the controversy over which is better — sugar or honey — but values of each will be discussed so that you can make your own decision. The important facts are that sweetening is important, and we should have some in our food storage. Besides, I haven't figured out how to make chocolate chip cookies without sugar.

> Sugar is the food from which the yeast plant makes leavening gas. Yeast makes carbon dioxide gas and alcohol from sugar. Yeast and sugar work together to form the tiny gas bubbles which permit the light porous texture of yeast products.
>
> Sugar adds flavor to the bread. It also helps its browning in the oven. Too little sugar prevents browning. Breads brown too quickly if too much sugar is used.
>
> White sugar is used most commonly in bread. Brown sugar, honey or molasses often are used in whole wheat or in fancy breads.[1]

Refined Sugar

Many forms of refined cane or beet sugar are available on the market today, ranging from granulated sugar to confectioners or powdered sugar. Sugar is valuable as an energy food.

> *Brown sugars* vary in color and flavor from the very light yellowish sugars to very dark brown ones. The lighter the color the higher the stage of purification and the less pronounced the

---

[1]Miller, *Yeast Breads,* p. 9.

flavor. Brown sugar retains some of the molasses from which the crystals are separated. . . .[2]

Raw sugar is the unrefined residue after removal of molasses from cane juice. It contains a fairly high proportion of some minerals, but like refined sugar is mainly carbohydrate.[3]

How Much to Buy

If you plan your buying from recipes, it pays to know how much sugar to buy. The following is a guide for quick reference:

- — One pound of granulated sugar yields about two and a half cups.
- — One pound confectioners powdered sugar yields about four cups if unsifted, or about four and a half cups if sifted.
- — One pound light or dark brown sugar, firmly packed, yields nearly two and one-third cups.

How to Store

Granulated, very fine, and powdered sugars absorb moisture and should be stored in a cool, very dry place. Remove sugar from the paper sacks or boxes it comes in and put it into an airtight container that will keep out moisture and prevent lumping. Sugar will store for years, if cared for properly. A five-gallon metal can will hold twenty-five pounds of sugar; a one-gallon glass jar will hold six to eight pounds.

If granulated sugar should get hard, try these suggestions. Cover it, set it in a warm place to dry out, then roll it with a rolling pin. If it turns very hard, like a brick, use it to make syrup for canning or freezing. If the sugar is still in a sealed bag, drop it flat side down onto the floor (keeping your toes out of the way). This will help to break it into usable chunks.

Brown sugar needs moisture. Place it in a tightly sealed container. Heavy plastic storage bags will serve for a short time, but not for extended storage. Gallon jars make good storage containers for brown sugar, as do the large plastic buckets with lids.

Some people put slices of bread or half an apple in the container to help keep brown sugar moist. Should brown sugar cake, put it in a heavy paper bag, wrap the bag in a damp cloth, and let it stand until soft.

---

[2]Hughes and Bennion, *Introductory Foods,* p. 276.

[3]USDA, *Yearbook of Agriculture,* p. 32.

If brown sugar becomes very hard, place it on a cookie sheet and put it in the oven at the lowest temperature for about thirty minutes. It will become soft enough to measure into convenient amounts such as a cup, or half a cup. As it cools, it will again become hard, but it can be used in cooking.

Hardened brown sugar can be used for making homemade syrup. Dissolve one cup of brown sugar in one cup of water, then boil it down to the desired thickness. Add maple flavoring if you wish.

Brown sugar, if properly stored, will keep for many years.

Syrups

Corn syrup, produced in the manufacture of corn sugar, sometimes contains added coloring and flavoring because by itself it has only sweetness. Corn syrup is used as the basic ingredient in many table syrups.

Maple syrup is the most popular syrup available on the market. It is the sap of the sugar maple tree, concentrated to the desired consistency.

The *sorghum* plant, like sugar cane, is a large coarse grass. After processing, sorghum is like molasses in appearance. It is a source for calcium and iron.

Molasses, a concentrated sweetening, is a product left after the sugar crystals have been removed from the sugar cane or beet juices. It is in demand chiefly for its flavor and the variety it provides. "The content of iron and calcium is high, especially in the darker kinds, but because molasses generally is used in small quantities it does not make an important contribution to the ordinary diet."[4] (This would be true of sorghum as well.)

Storing Syrups

Syrups store well, but for convenience in handling it might be wise to purchase syrup in a small container, such as a five-pound can. If kept in an airtight container in a cool place, syrup should keep for several years.

If sugar crystals form in syrup, it can be heated until the crystals dissolve. This does not harm the syrup.

---

[4]USDA, *Yearbook of Agriculture,* p. 33.

Honey

"Isn't it funny how a bear likes honey," says Winnie the Pooh. It isn't just Pooh bears that like honey, for honey is one of the oldest known sweets. Honey is the nectar of flowers, gathered, changed and concentrated by honeybees.

Many exaggerated claims have been made about honey. If a person is diabetic, honey has an advantage over sugar in the diet. Many questions have been raised about the nutritional value of honey versus the nutritional value of sugars. On this, refer to the accompanying table.

> Medium dark molasses contains enough calcium and iron that a cupful would supply the daily requirements of both. Brown sugar contains about one-third as much calcium and half as much iron as molasses. White sugar and honey contain practically no other nutrients than carbohydrates. Honey is much more expensive than sugar. Honey is not a perfect food, as no amount will furnish all the necessary nutrients. It is easily digested, as is sugar. Honey is a good sweetener and may contain more trace minerals than white sugar. As far as vitamin C is concerned it does not contribute significantly to the vitamin C requirements.[5]

It clearly becomes a matter of personal preference and individual taste.

The flavor of honey varies according to the flowers from which the bees gather nectar. It can be mild or strong. As a rule, the lighter the color of the honey, the milder the flavor.

Kinds of Honey

Six types of honey are on the market today: Liquid; granulated or solid (sometimes called candied); creamed; comb; cut comb, which comes in bars about four inches long and one and a half inches wide; and chunk, with small pieces of honeycomb in the liquid honey.

Cooking with Honey

You will get best results in cakes and cookies made with honey if you use recipes developed with honey as an ingredient. If you have not cooked with honey, but are considering storing some, practice cooking with it now. It is not difficult to use, just different.

---

[5]*Home Storage Handout* (Salt Lake City: Relief Society, 1974).

NUTRITIVE VALUE OF SUGARS

Dashes in the columns for nutrients show that no suitable value could be found although there is reason to believe that a measurable amount of the nutrient may be present.

Food, Approximate Measure, and Weight (In Grams)		Water	Food Energy	Protein	Fat	Fatty Acids			Carbohydrate	Calcium	Iron	Vitamin A	Thiamine	Riboflavin	Niacin	Vitamin C
						Saturated	Unsaturated									
							Oleic	Lin oleic								
	Grams	Per Cent	Calories	Grams	Grams		Grams	Grams	Grams	Milligrams	Milligrams	International Units	Milligrams	Milligrams	Milligrams	Milligrams
Honey — 1 Tbsp.	21	17	65	trace	0	—	—	—	17	1	.1	0	trace	.01	.1	trace
Molasses, cane																
Light — 1 Tbsp.	20	24	50	—	—	—	—	—	13	33	.9	—	.01	.01	trace	—
Blackstrap — 1 Tbsp.	20	24	45	—	—	—	—	—	11	137	3.2	—	.02	.04	.4	—
Sorghum — 1 Tbsp.	21	23	55	—	—	—	—	—	14	35	2.6	—	—	.02	trace	—
Table blends, chiefly corn, light or dark																
1 Tbsp.	21	24	60	0	0	—	—	—	15	9	.3	0	0	0	0	0
Sugars																
Brown, firm packed																
1 cup	220	2	820	0	0	—	—	—	212	187	7.5	0	.02	.07	.4	0
White																
Granulated																
1 cup	200	trace	770	0	0	—	—	—	199	0	.2	0	0	0	0	0
1 Tbsp.	11	trace	40	0	0	—	—	—	11	0	trace	0	0	0	0	0
Powdered,																
Stirred before measuring																
1 cup	120	trace	460	0	0	—	—	—	119	0	.1	0	0	0	0	0

USDA, *Nutritive Value of Foods,* Home and Garden Bulletin, No. 72.

Honey can be used measure for measure in place of sugar in preparing puddings, custards, pie fillings, baked apples, candied or sweet-sour vegetables, and salad dressings.

Cakes, cookies and breads made from honey remain moist during storage. Crisp cookies are likely to lose their crispness if they stand too long.

In cakes, honey can replace as much as one-half of the sugar without making it necessary to change the proportions of the other ingredients.

With cookies, the type of cookie determines how much sugar can be replaced with honey. For gingersnaps, replace no more than one-third of the sugar with honey. In brownies, honey may be used for half of the amount of sugar. In fruit bars, use up to two-thirds the amount of sugar called for.

A rule of thumb on substituting honey for sugar is to reduce the liquid by one-fourth cup for each cup of honey used to replace the sugar. When honey is substituted in baked goods, add one-half teaspoon baking soda to the mixture for each cup of honey used, and bake at approximately twenty-five degrees lower temperature.

To prevent a soggy layer of honey forming on the top of cakes and cookies, be sure to mix it thoroughly with the other recipe ingredients. Combine the honey with shortening or other liquids.

When baking with honey, if you measure the shortening first, then measure the honey in the same cup, the honey will slide out easily.

It is wise to use a mild-flavored honey in cooking. Honey that is strong-flavored could easily overpower the other flavors in a recipe.

Ways to Serve Honey

— As a sweetener on hot or cold cereals.

— To sweeten lemonade, milk drinks or eggnog.

— To sweeten whipped cream for a dessert topping.

— To drizzle over fresh fruit, fruit salad or ice cream.

— For serving on waffles, pancakes or toast.

— In honey butter, by mixing equal parts of margarine or butter and honey.

— As sweetener in mashed sweet potatoes or yams.

— For a fruit salad dressing. (Mix ¼ cup honey with ¾ cup sour cream.)

Simple Candy Recipe

Cooked to the hard-crack stage and stretched like taffy, with vanilla added, honey makes a delicious old-fashioned candy.

Tips on Storage

Honey should be stored in fairly small containers with tight-fitting lids, since honey absorbs and retains moisture and loses its flavor and aroma when exposed to air. Refrigeration or freezing won't harm the flavor or color, but it will hasten granulation. If honey crystallizes, place the container in a pan of warm to hot water until the crystals dissolve, but *do not cook the honey.* (This is a good reason for storing honey in small containers. It is rather difficult to place a five-gallon can, weighing sixty pounds, in a pan of hot water on the kitchen stove.) Crystals do not affect the food value or flavor of the honey. After honey has been stored for a long time it tends to get darker, but this does not harm it.

Substitution in Recipes

It usually pays to use the exact ingredients called for in a recipe, but if you suddenly find your sugar or honey can empty you can substitute other sweeteners.

When you need	Use
1 cup granulated sugar	1 cup firmly packed brown sugar
1 cup granulated sugar	1 cup molasses, syrup or honey and ¼ to ½ teaspoon baking soda, and ¼ cup *less* liquid
1 cup granulated sugar	1½ cups maple syrup and ¼ cup *less* liquid
1 cup corn syrup	1 cup granulated sugar plus ¼ cup liquid. (Substitute for only half the sugar)
1 cup honey	1 cup molasses and omit baking soda. Replace each ¼ teaspoon soda with 1 teaspoon baking powder
½ cup honey	½ cup granulated sugar
1 cup honey	1 cup sugar plus ¼ cup liquid

22
Yeast

It is not known when leaven was first used to lighten bread, but the Egyptians are given credit for the first accounts of it. The story is told that Joseph probably ate unleavened bread in Shechem, but ate leavened bread in an Egyptian prison. The Greeks and Romans made bread with leaven. They prepared a ferment by making a batter of bran and fermenting cider and spreading it out in the sun until it was dry. It could be kept any length of time. When ready for use, it was soaked in water and mixed with flour.

When methods other than fermentation were first used for leavening is unknown.

> Yeast breads are so called because they are leavened or made light by the action of microscopic round or oval one-celled plants called yeasts. Certain strains of wild yeast plants have been cultivated and are sold as dry or moist compressed yeast cakes. Each cake consists of millions of these tiny plants pressed into a solid mass.[1]

The main purpose of yeast is to raise dough. Yeast is responsible for the bubbles of gas which cause the dough to rise.

The time may come when it will be necessary for you to do your own baking. Don't wait until then to find out the value of yeast. Why not bake and enjoy bread or rolls now. The only problem may come after you practice for a while and then try to go back to using "store bought" bread.

Become familiar with the several forms of yeast available, then choose the kind that will work best for you.

---

[1]Fern Silver, *Food and Nutrition* (New York: D. Appleton-Century Co., 1941), p. 366.

Compressed yeast, bought in cake form, is grayish in color, brittle, and crumbles easily when pressed between the fingers. It is quite perishable, but will keep in the refrigerator for about two weeks if tightly wrapped. When stale, yeast becomes slimy and brown in color, with a strong odor. A dark brown color usually means that the yeast spores have died and it will not cause dough to rise properly.

Active dry yeast comes in a fine dried granular form. It contains much less moisture than does the compressed cake, and is thus much less perishable. It can be purchased in small foil envelopes and will keep for several weeks without refrigeration, but keeps much longer if stored in a very cool place. Dry yeast is also available in a two-pound can in many markets. In this type of container it will keep for eighteen months to two years, unopened, if stored in a cool place. It can be frozen if you have room in the freezer. Once it has been opened, it should be kept tightly covered in the refrigerator until it is used.

Interchange Yeasts for Baking

If your recipe calls for a yeast cake, how much dry yeast should you use? You can use them interchangeably — one envelope dry yeast (about 1 Tbsp.) for one cake of yeast (½ ounce). The dry yeast must be dissolved in warm water. Reduce the total amount of liquid used in the recipe accordingly.

Starters or Sourdough

A starter or sourdough is a form of yeast that can be made and kept at home. Sourdough starters can be "started" by mixing yeast with equal amounts of milk or water and flour and allowing this to be exposed to the air so that the yeast will grow. After two to five days it should be sour and bubbly and ready to use.

When the sourdough mixture is used in baking, be sure to save at least a cupful as a start.

A special crock to keep your yeast in is not necessary. Any covered plastic or glass container will serve.

Sourdough must be "fed" if it stands for any length of time between bakings. If replenished or "fed" with flour and water every seven to ten days, and allowed to stand at room temperature until bubbly and fermented, the starter will keep indefinitely. If it is not used within ten days, add one teaspoon of sugar. Some starters are reputed to date back several generations.

Simple Sourdough Starter

2 cups flour
1 Tbsp. dry yeast or 1 cake yeast
2 cups warm water

Combine the ingredients and blend them very well. Put the starter in a warm place and allow it to stand overnight, all day, or even several days. When you are ready to use it, put half of the starter in a covered jar and store it in the refrigerator. If it is replenished every week with flour and water or milk, the starter will keep a long time. Prior to using the starter, allow it to stand for several hours at room temperature.

The longer a start is kept and used, the more "sour" the dough will become.

Liquid Yeast

Liquid yeast is another homemade yeast. It is usually made from potato water, sugar, and yeast. Like a sourdough starter, it must be used regularly to keep it from spoiling.

Liquid or Everlasting Yeast

1 quart warm potato water
½ Tbsp. dry yeast or
½ cake yeast
2 Tbsp. flour
1 tsp. salt
2 tsp. sugar

Mix together all of the ingredients, and put it in a warm place to develop. Keep half of the amount for a "start" next time. The next time you want to use it, add the same ingredients, except the yeast, and allow it to stand at room temperature for several hours. Between uses, keep it in a covered jar in the refrigerator.

Always allow sourdough or liquid yeast to stand at room temperature for at least one hour prior to using them in baking. This allows the yeast spores to become active.

One cup of yeast per batch of baked goods is usually adequate. Reduce the liquid called for in the recipe by the amount of liquid yeast used.

Begin to experiment with various yeasts now to find one that suits your needs. Don't say I didn't warn you about homemade bread "problems" though. Even if the yeast fails and the bread raises only a little, if it's homemade, it will be eaten!

23
Canned Goods

> The history of the canning industry goes back to the year 1795, when the French government sought to improve the food supply of its army and navy by offering a prize for the discovery of some method of preserving food from one harvest to another. A French candymaker by the name of Appert worked on the problem for years. Finally in 1809 he succeeded in preserving certain foods by sealing them in especially made glass bottles in which the foods were boiled. He founded a commercial food preserving firm which is still in business today. Thus modern commercial canning as well as home canning was started.[1]

Canned goods are the major item in many pantries. We rely on a can of this and a can of that to make a quick supper. During the winter months, many of the fruits and vegetables we use are canned. Canned meats are a reliable convenience. So much for ordinary use. Now let's focus on canned goods as an excellent way to have a stock of meat, fruits and vegetables on hand for use in any emergency.

Proper Storage Conditions

Proper storage conditions almost guarantee a good shelf life for canned goods. The length of time canned goods can be stored is determined chiefly by the storage temperature. A dry place with a moderately cool but not freezing temperature (50° to 58° F.) is ideal. Quality loss at 85° F. is about twice that for an item stored for the same length of time at 67° F. The cooler and dryer canned goods are kept, the longer they will last. Under proper storage conditions, most canned goods remain usable for several years.

[1]Silver, *Food and Nutrition,* p. 455.

Only high quality canned goods should be considered for storage. Food in glass containers should be kept in a dark place.

Place older cans at the front of the shelves and use them first. Make it a habit to label and date everything in order to avoid waste caused by too long storage. Small strips of adhesive tape and a pen (felt tip laundry marker) with dark ink will do the job. A wax pencil works too.

Do not store cans or containers directly on a cement floor, as cement has a tendency to sweat, and this will cause the cans to rust and deteriorate. Avoid a moist storage area for canned goods. Attics, kitchen cabinets, or garages without insulation are poor storage areas because they lack temperature or humidity control.

Shelf Life

When the length of their shelf life is considered, canned goods fall into several groups. Products high in acid have a short shelf life. This includes grapefruit and orange juice, black and red cherries, berries, prunes and plums. These foods store well for one to two years. Fruits such as peaches, pears, apricots, and applesauce generally keep two to three years. Canned vegetables such as beets, carrots, green beans, spinach, tomatoes and tomato juice should keep from three to four years. Vegetables and meats such as peas, corn, lima beans and roast beef generally have a shelf life of three to five years. Canned milk should be agitated about every thirty days, and should be used within one year. It is best to have a regular turnover at least once a year.

Nutritional Value

The nutritional value of canned goods is no doubt subject to many factors, including any losses due to the high temperatures used in the canning process. There may also be losses in nutrient values during shelf life.

> The greatest factor to speed up these changes is the temperature at which food is stored. The reactions in fruits and vegetables are approximately doubled with each 18° F. increase in temperature. These reactions bring about changes in the flavor, color, texture and nutritional value of the canned product.[2]

There are those who suggest that due to methods of processing, storage, etc., there is virtually no food value left in canned goods.

---

[2]*Food Storage in the Home* (Logan, Utah: Utah State University Extension Service, 1973), p. 8.

I know of no evidence for this. Granted, there are fewer vitamins in a can of carrots than in a bunch of carrots just pulled from the garden, but it is still acceptable food.

Canned goods are usually safe to eat as long as the seal of the can is not broken. *Do not* eat the contents of a can which is leaking or has a bulging end. When a can is opened, if there is a spurting liquid, an odd odor, or mold on the food, throw it out without tasting. Don't take chances.

Emergency Cooking

If there was a power outage for many hours or even several days, a can of soup or pork and beans could be heated over a sterno stove or canned heat or even a barbeque.

When heating canned food in the can, always remove the lid to prevent a pressure build-up and possible explosion. Place the can or jar in hot water and simmer. If no water is available for such "luxury" uses, place the can directly over the heat, watching it closely to avoid scorching. Use a folded towel or hot pad to hold the can, and stir it with a fork.

One advantage to cooking in a can in an emergency, when there might be power or water shortages — water or heat wouldn't be wasted doing unnecessary dishes. (Funny how the younger members in my family keep hoping for power shortages and no water!)

If you have children in your family, they will enjoy practicing having "hobo" suppers or lunches. The ages of children who enjoy this kind of thing usually range from five to ninety-five.

Hints on Buying Canned Goods

Canned fruits come packed in light, heavy or extra heavy syrups, in water, in slightly sweetened water, or in fruit juices. The heavier the syrup, the sweeter the fruit and sometimes the higher the price. Special diet packs are the most expensive. You may wish to buy more than one style pack, according to the intended use.

Whole vegetables usually cost more than cut styles, because it is hard to keep such fragile products whole during processing. Short cuts, diced and pieces are the least expensive.

Is a Sale a Sale?

Be aware that a sale is not always a sale. Items may be placed on sale as "loss leaders" because the stock is old. Even "old" loss leader sales are a good buy if you use the items constantly and such items would not be stored too long. For instance, at our house the catsup is lucky to make it from the bag to the shelf — no storage problems there.

To keep foods moving, grocers usually run at least three or four promotions a year on canned goods. Watch for these. Take time to read the labels. Nationally known Brand X is being promoted at so many cans for a dollar, and this could be a good price for that brand. Take a few minutes to compare. Many of the supermarket chains now have their own brands. Compare ingredients, quantity, quality, and cost of the sale item with the store's own brand of that product. It might prove not to be a bargain after all. A display of food placed at the front of the store or at the end of the aisles with large bold signs proclaiming "sale" does not always mean there is a discount in price for the consumer.

Some people claim that it does not pay to "shop the ads," that it uses more gas than can be saved by running between stores. If you go about it purposefully, you can save a great deal of money. But first, become and stay familiar with current prices. This may mean spending half an hour to forty-five minutes studying the sale papers. Unless you know "regular" prices, how can you recognize "bargain" prices? The ideal situation is not to have to buy anything unless it is on sale — then you save money.

As you read the ads and prepare to shop, do so with a pen and paper. As you study each new ad, write down the store name, then list the items that are of value to you, the prices, and how much you will spend for that item.

Sale Items	Amount to spend per item
Smith's Grocery	
hamburger — 79 cents per lb.	$3.00
laundry soap — lb. box $1.10	2/2.20
catsup — lg. size — 3/$1.00	3.00
Jones' Store	
peaches — lug $2.00	4.00
bananas — 5¢ lb.	2.00
Rx Drugstore	
deodorant — 3/$1.00	2.00
shampoo — 79¢	.79
aspirin — 19¢/100	????

If you have previously decided on a set amount to spend for storage, you know exactly what is within your limits. If you are like most of us and this amount must come out of the regular grocery budget, by checking the list you can immediately determine how much you should spend. If you will discipline yourself to purchase only the items on the list in the specified amount, you will save money. Promotional sales are offered to get you to come into the store and buy the sale items *and* many other things which are not on sale. If you want to save money, take advantage of the sale items, and ignore everything else.

An article appearing in the *Deseret News* was entitled, "Do Food Specials Save You Money? You Bet!" A survey conducted by the newspaper indicated at least a 27 percent saving.

Some have suggested that the high cost of gas prevents you from saving money, but there are ways to save money by shopping the ads. A car pool might be the answer. (It would be nice if one member owned a station wagon.) Others in your neighborhood besides yourself are interested in saving money. Plan your route between stores so that it is orderly, and you won't waste gas. Start at one side of the shopping area, working toward the other. "Where there's a will there's a way." Besides, it's actually fun to see if you can win a battle or two in the war with prices.

24
Emergency Water Supply

Next to oxygen, water is the most essential element in the survival of men and animals. A person can do without food for several weeks or more, but without water he can survive for only a few days.

Water is not usually thought of as a food because it does not provide energy for the body but it is an essential part of all body tissue. Most water contains small amounts of valuable minerals unless they have been removed by a water softening process.

You should consider water an essential item in your pantry. It isn't practical for most of us, especially if we are city dwellers, to have a large reserve of water. Discussed in this chapter are a few ways to store water for an emergency situation only; for instance, in the event that a city should have problems with the water treatment system or a severe storm should damage the pipes.

The Civil Defense Preparedness Agency recommends that we have on hand at all times a two-week supply of water. On an emergency basis only, that would be seven gallons of water per person. This amount is calculated on the basis of one-half gallon a day for each person for drinking and food preparation. If water is desired for other uses such as bathing, brushing teeth, washing dishes, etc., another seven gallons per person is recommended for this purpose.

If you have on your shelves cans or jars of fruits, fruit juices, vegetable juices, or soups, supplemental liquid would be available from this source. Canned tomatoes can easily be made into soup or juice.

Ways to Store

Reserve water may be stored in a number of ways. Thoroughly washed, clean plastic jugs with tight-fitting lids, or glass jars or bottles with screw caps can be used.

Metal containers often give the water an unpleasant taste and are prone to rust. Plastic has an advantage over glass by being shatterproof if it should fall or be bumped. Clean fruit jars can be filled with water. Leaving one inch of headspace, process the jars in a boiling water bath in the same way as fruit juice. Process quart jars for twenty minutes; two-quart jars for twenty-five minutes.

Clean water is also available in your hot water tank and toilet tanks. Water can be obtained from the hot water tank by opening the drain faucet at the bottom of the tank. To get the water to flow, you may have to turn on a faucet somewhere on the water line. This creates pressure in the line and allows the water to drain out of the hot water heater. There are twenty to sixty gallons of water in a hot water heater, depending on its size. A toilet flush tank holds approximately five gallons of water.

If you have camping or picnic equipment, keep the jugs and other containers filled with water between outings.

Clean Water

It is important to store water that is clean and uncontaminated. Water that has been tested by health authorities and found safe would be safe to store. This means that water from the tap ordinarily does not need to be treated in any way.

If there is any question about the safety or cleanliness of the water you intend to use or store, purify it.

How to Purify Water

Boiling. The safest method for purifying water is to boil it vigorously for three to five minutes in order to destroy bacteria that might be present. To improve the taste of water after it has been boiled, pour it from one clean container to another, several times. This will put the air back into the water.

Easy bleach method.

> Any household bleach solution that contains hypochlorite, a chlorine compound, as its only active ingredient will purify water easily and inexpensively. [Read the label *before* you use it.]

Bleach solutions with 5.25 percent of sodium hypochlorite are most common. They are available in grocery stores. Add the bleach solution to the water in a clean container in which it can be thoroughly mixed by stirring or shaking. The following table shows the proper amount of a 5.25 percent solution to add to water. . . .

Amount of Water	Amount of solution to add to Clear Water	Cloudy Water
1 quart (¼ gallon)	2 drops	4 drops
1 gallon	8 drops	16 drops
5 gallons	½ teaspoon	1 teaspoon[1]

Add the chlorine solution to the water and stir, then allow the mixture to stand for thirty minutes. The water then should have the distinct smell or taste of chlorine. If this taste or smell is not present, add another dose of solution to the water and let it stand another fifteen minutes. The taste or smell of chlorine in water treated this way is a sign of safety.

Other Methods. A 2 percent tincture of iodine can be used to purify small quantities of water. Add three drops of iodine to each quart of clear water, six drops to each quart of cloudy water. For a gallon, add twelve drops for clear water, twenty-four drops for cloudy water. Stir well.

Inexpensive water purification tablets are available at most sporting goods stores. Use these tablets as directed on the package.

Making a Filter

Though a filter strains water in a mechanical way and does not remove or destroy bacteria, knowing how to filter water might be helpful in an emergency.

Using a wooden tub or barrel, or a large plastic bucket, make a tight false bottom two or three inches from the bottom. Perforate this with small holes, and cover it with a piece of clean canvas. Cover this with a layer of pea gravel several inches deep. Add a layer of clean, washed, coarse sand two to three inches deep, then add coarsely granulated charcoal (about the size of peas) until the

[1]USDA, *Family Food Stockpile For Survival,* Home and Garden Bulletin No. 77 (Washington, D.C.: The U.S. Government Printing Office, 1966), p. 11.

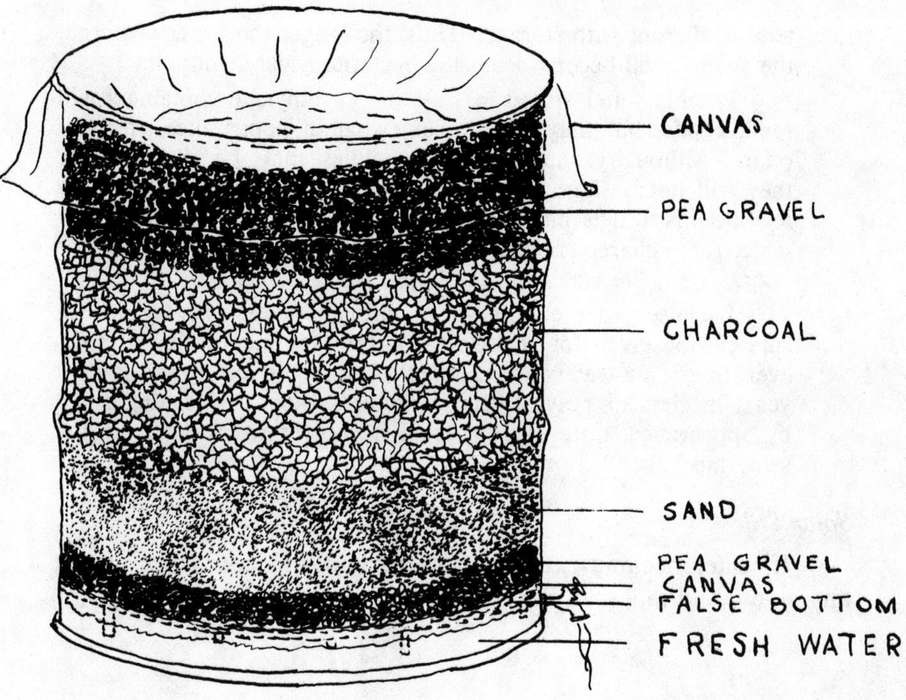

Homemade water filter

container is filled to within two or three inches of the top. Add another layer of pea gravel, and put a piece of canvas over the top as a strainer.

If a very large container is used, a spigot should be installed in the bottom. If a smaller bucket is used, make one or two holes in the bottom of the bucket and place it over a container to catch the water.

Filtering can be done on a large or small scale, depending on how much water you need. A very small, inexpensive filter can be made from a flower pot. Insert a small sponge in the drain hole and fill the pot as directed for the larger filter, then place it on top of a jar, which will catch the water.

Shelf Life of Water

If stored in clean containers and if safe bacterially at the time of storage, water will remain safe because disease organisms tend to die out with storage. Thus, the longer the water is stored, the safer it will become from the bacteriological standpoint.

Potable water stored in glass or polyethylene containers will remain safe, but may change somewhat in appearance, taste or odor. Although some of these qualities may be disagreeable, they will not be harmful. Stored water should be checked every few months to determine whether containers have leaked or if any undesirable characteristics have developed in appearance, taste or odor. If so, the water may be replaced.

Because water quality varies throughout the country, no set rule can be given for shelf life. Current experience shows, however, that some waters taken directly from a tap and stored several years in glass or polyethylene containers cannot be distinguished by appearance, taste or odor from freshly drawn water from the same tap.[2]

Solar Still

If water became extremely scarce and you could not obtain any, you could make a solar still.[3]

---

[2]*Facts on Water Storage for Emergency Use* (Salt Lake City: Utah Office of Emergency Services, 1974).

[3]See Boy Scouts of America *Fieldbook for Men and Boys,* p. 315, or Esther Dickey's *Passport to Survival* (Salt Lake City, Utah: Bookcraft, 1969), p. 129.

25
Salt, Spices, Herbs and Seasonings

Salt

Salt may be considered indispensable. Wild animals sometimes travel miles to a "salt lick" to satisfy their craving for it. Salt is said to have over fifteen hundred uses, one of which is to bring out the flavor of other ingredients.

> It aids in the coagulation of protein. Salt lowers the freezing point of water and is added to ice for freezing ice cream and other mixtures. . . . Salt also raises the boiling point of water solutions. Salt controls the fermentation of yeast to make baked goods more desirable. Salt in high concentration is used in preserving foods to retard bacterial action.[1]

It is also used in curing meats and fish, and in the making of pickles, butter and cheese. Even the taste of some sweet foods can be improved by adding a few grains of salt. Salt should not be added to the liquid in which yeast is softened, as it may retard the action of the yeast.

Salt, like sugar, can be stored almost indefinitely. Unless it is kept free of moisture, it will pack and lump. Usually the heavy one-pound cardboard containers protect salt adequately, but if it is purchased in a sack, transfer it to a more suitable container.

A supply of salt is good to have on hand. Can you imagine popcorn or corn on the cob without salt?

Spices and Herbs

Spices and herbs, among the oldest products known to man, have a colorful history. Countries have been discovered as a result

---

[1]USDA, *Yearbook of Agriculture*, p. 35.

of the search for precious spices. Historical records give much evidence of the use of herbs in early times. Many of the ancient philosophers wrote of herbs, and the Bible mentions herbs and spices.

People of many lands have used spices and herbs and still use them today to give variety and interest to family meals. No matter what foods you keep in your pantry, good flavorings are essential. It would be impossible to make boiled wheat taste like spaghetti without bay leaves, oregano, thyme, garlic and tomato sauce. How could you make chili without chili powder or cayenne pepper?

A time of hardship is not a time to try out new tastes and new foods, so keep on hand a good supply of spices and herbs to give the same good taste to all of your favorite foods.

> Generally herbs are leafy aromatic plants, usually grown from seed in the temperature zone. Spices are often pungent barks of trees, or seeds, or buds of plants grown in the tropics. Some seasonings, such as mustard, start out as herbs, and after going to seed are spices.[2]

Amounts to Use

Since the pungency of each spice varies, no blanket rule can be given for the amount to use.

> Where no recipe is available, start with about ¼ teaspoon of spice to each pound of meat or pint of sauce or soup. Use ⅛ teaspoon in the case of red pepper (or cayenne) and garlic powder. You will probably find these starter quantities should be increased in many instances, but it is easier to add than subtract.[3]

Since dried herbs are about three times as strong as fresh, fresh herbs can be used in greater quantity than the dried.

When to Add Them

Because ground spices tend to give up their flavors quickly, they should be added near the end of the cooking time, when you add salt and pepper. Whole spices, excellent for long-cooking dishes, should be added at the beginning of cooking. Crumble whole herbs just before they are added.

---

[2]Frederic Rosengarten, Jr., *The Book of Spices* (Livingston Publishing Co., 1969).

[3]*A Guide to Spices,* Technical Bulletin 190, 2nd Revision (Englewood Cliffs, N.J.: American Spice Trade Association, Inc.), p. 2.

How to Store

Store spices in as cool and dry a place as possible. Since both heat and light cause spices to lose flavor, never store them above or close to the stove. Storing them in tightly closed containers will prevent evaporation of their oils and aromas. Whole spices keep longer than ground ones. Herbs tend to lose their flavor a little faster than some spices.

For economy and better keeping quality buy in large metal cans the spices you use most frequently, then transfer small amounts to a jar in your spice cabinet as needed. Store the large cans in a cool, dry place.

Tips on Herbs and Spices

Cook with one herb or spice at a time until you know its characteristics.

The greener the herb or the redder the spice, the more likely it is to be fresh and have a good flavor.

To tell if spices are fresh, open the container and sample the aroma. If it does not have a strong, pungent odor, it is probably stale and should be replaced.

How mild or how strong is it? Crush a small amount in the palm of your hand, allow it to become warm, then sniff it.

The secret of cooking with spices is to use only enough to enhance the natural food flavor, not dominate or overpower it.

Part II
NONFOODS

26
Fuels and Heating

When plans are made to meet emergencies, fuel and heat are all too often overlooked. So excellent are the services of the electric and natural gas companies that we think nothing could happen to them. But they are subject to breakdown, fuel shortages, and natural disasters. Some ideas are here presented for providing yourself with fuel and heat if the old standbys should falter. You will want to prepare now while there is time. As you study these ideas, be creative and adapt them to your own circumstances.

Kinds of Fuel

Coal

Coal stores well if kept dark and away from air. Air speeds deterioration and breakdown, causing it to burn more rapidly.

Coal can be stored in any of these ways:

- — Excavate a pit and store the coal in bags, barrels, or bulk. The pit should be covered.
- — If it is stored in bulk, line the pit with plastic to help keep out air and to help in recovering the coal for use.
- — Store in bulk in a basement room.
- — To store above ground, place it on the north side of buildings to keep it out of the sun as much as possible. Cover it and protect it from the weather.

The amount of coal needed for heat will depend on the home size, the amount of insulation, and the temperature to be maintained. A house with a thousand square feet of space, three inches of wall insulation, and six inches of ceiling insulation, if kept at 70° F., would use approximately five to six tons of coal in one year.

The cost of coal varies from company to company, depending on the time of year, whether or not it's delivered, and other factors. It pays to compare prices.

Wood

It takes approximately eighteen cords of the best hardwood to heat the average home for one year, making it about the most expensive fuel that can be used for home heating. If soft wood is used, figure on burning more.

Hardwood, which is slow burning and sustains coals, feels solid and heavy in the hand. Because hardwood is more difficult to burn than soft wood, a good supply of kindling is necessary to get it going. Soft wood is lightweight and burns very rapidly. leaving a good supply of ashes, but very few coals for cooking.

When gathering firewood, consider these points:

1. Rotten wood crumbles, smolders, smokes and gives off very little heat.
2. Dry branches and twigs make excellent tinder.
3. Green branches and twigs bend easily and make very poor fuel.
4. Logs with soft or pithy centers do not burn well.
5. Split logs burn more easily than smooth, round ones.
6. Waterlogged wood is difficult to burn.

Newspapers for Fuel

Use the following method to make newspaper logs, a good, inexpensive source of fuel.

1. Take about eight pages from your newspaper and open them out flat. Alternate the cut sides with folded ones and slide them together (should be one page width). Place a one-inch wood dowel or metal rod across one end and roll the paper around the rod very tightly. When there are six to eight inches left to roll onto the rod, slip another eight pages underneath the roll. Continue this procedure until you have a roll four to six inches in diameter. With a fine wire, tie the roll on both ends, withdraw the rod and your newspaper log is ready to use. Four of these logs will burn for about one hour.

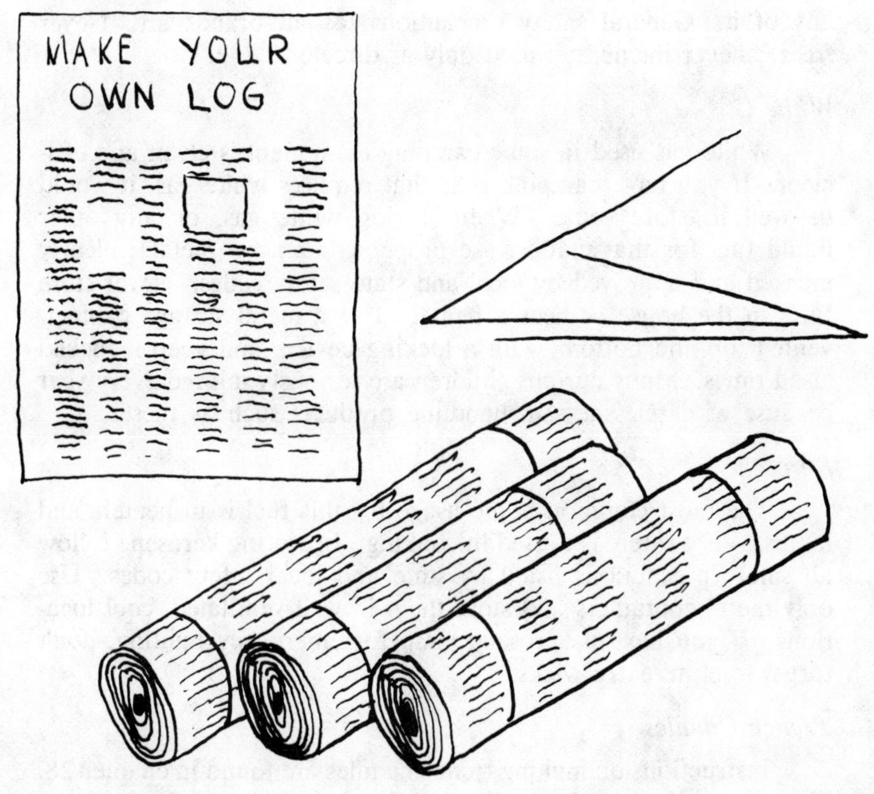

Newspaper logs for fuel

The average weekday newspaper should make one to two logs. Pound for pound, newspaper logs are about as efficient as wood as an energy source.

Propane

Propane can be purchased prepackaged in throwaway containers or can be stored in your own containers. This is one of the cleanest, most efficient fuels available. If you have camping equipment that is designed for propane, store a few containers of fuel. Your camping equipment is ideal for emergency use.

There are many different brands of propane on the market. Thoroughly read the instructions on use and disposal before using any of it. General safety precautions for all brands are: Never freeze; never incinerate; used only as directed.

White Gas

White gas used in some camping equipment is clean and efficient. If you have camping gear that requires white gas, it would be well to store some. When storing white gas, or any other liquid fuel for that matter, use proper containers (metal), clearly marked and approved by local and state safety codes. Never store fuels in the house or near a heater. Use a metal storage cabinet, vented top and bottom, with a locking device, and keep it locked at all times. Many curious children are seriously injured every year because of carelessness in handling products such as these.

Kerosene

The most common home usage for this fuel is in heaters and lamps. Very rarely is it used in cooking. In storing kerosene follow all safety precautions listed by state and local safety codes. Use only metal containers and store them in well-ventilated, cool locations. If you have a kerosene lamp for emergency lighting, don't forget to store extra wicks.

Trench Candles

Instructions on making trench candles are found in chapter 28. Trench candles can be used as a fireplace fuel, much the same as a paper log. The difference is that trench candles are dipped in paraffin, and will burn longer than a simple paper log. When using them as fireplace logs, do not cut them into short sections as you would for candles.

Charcoal

With a few charcoal briquets, you can not only cook a meal but provide some heat. Available at nearly all markets, charcoal also can be made at home. It can be stored safely as purchased. Just keep it dry. Do *not* burn briquets in a nonventilated area.

For making your own charcoal, select twigs, limbs and branches of fruit, nut and other hardwood trees. Black walnuts and peach or apricot pits also make excellent briquets. Any of the foregoing will make a hot fire and give off very little smoke. After cutting the wood into desired sizes, place it in a can which has a few holes punched in it, put a lid on the can, and cook the briquets in a hot fire. Holes in the can allow gases and flames to escape, while exclusion of oxygen keeps the wood, nuts or pits from being completely consumed and turning to ashes. When the flames from the holes in the can turn yellow-red, remove the can from the fire, and allow it to cool. Store briquets in moistureproof containers.

SOME FUEL CONSUMPTION RATES

Fuel	Amount or size	Burning Time	Comments
White gas, lanterns —			
two mantle	2 pints	10 to 12 hrs.	
single mantle	2 pints	16 to 18 hrs.	
Kerosene lanterns	1 quart	45 hours	One wick lantern
Candles	¾″ x 4″	2 hrs. 20 min.	
	⅞″ x 4″	5 hrs.	
	2″ sq. x 9″ tall	63 hours	Based on 4½ hours burning time per day
Heaters	5 quarts	18 to 20 hrs.	8,000 BTU
Catalytic	3 quarts	12 hrs.	5,000 BTU
heater (white gas)	2 quarts	18 to 20 hrs.	3,500 BTU
Stoves	3½ pint		
two-burner	aerosol can	4 hrs.	Using both burners
	white gas	4 hrs.	Using standard setting
	20-lb. trailer tank	120 hrs.	stove
		200 hrs.	lantern

HEATERS

Catalytic Heaters

Several manufacturers of camping equipment make catalytic heaters which use white gas or propane. They store easily and, if taken care of properly, will give many years of service. The large ones will produce as high as 8,000 BTU of heat. For safety purposes, the area should be well ventilated whenever a catalytic heater is used inside a tent, room or other enclosed area.

Small Iron Stoves

Small iron stoves are great for heating or cooking, yet inexpensive to buy. Available with or without ovens, they could be a life-saver during an emergency. The stove surface can get very hot, so keep children away when these stoves are in use. For fuel, use wood, paper logs, charcoal or coal.

Barrel Heater

The barrel heater can be made at home very easily, and it could also be used to cook on. Any of the dry fuels — wood, coal, newspaper logs or trench candles — are suitable for this heater.

For making a barrel heater you need a fifty-gallon drum to serve as the heating chamber. Cut a 9-x-12-inch door in the bottom front where the large plug is situated. Two hinges and a simple latch, purchased at a hardware store, can be installed easily. On one side of the door measure two inches up from the bottom and two inches down from the top and mount the hinges. Mount the latch at the center of the opposite edge. In the back, top, flat surface, mark a circle the diameter of the stove pipe that will be used as an exhaust. Find the center of the circle, then drill a hole at this point large enough to insert a saber saw blade. Make eight straight cuts from the center to the outside edge of the marked circle, thus dividing the circle into eight pie-shaped pieces. (Do not remove them.) Bend each of the pie-shaped pieces up, hinging at the heel. Bend each piece a little beyond the vertical, so that you can wedge the stovepipe over them tightly.

For legs, cut four 1-x-10-x-¼-inch pieces of black strap iron and bolt them with ⅜-inch bolts on all four sides of the drum. The legs should fit four inches up the side of the drum. Two bolts through each leg and drum will do. There should be six inches of leg extending below the drum. A drill, bit, saber saw and blade can be rented at most equipment rental stores.

Now you are ready to light a fire and try it out. *Caution:* Select a barrel that is clean, one that has never had flammable material stored in it. Keep children away or they could receive severe burns.

If you use a barrel heater indoors, vent your exhaust through an existing flue or out a window. If the exhaust pipe is to go

Stoves

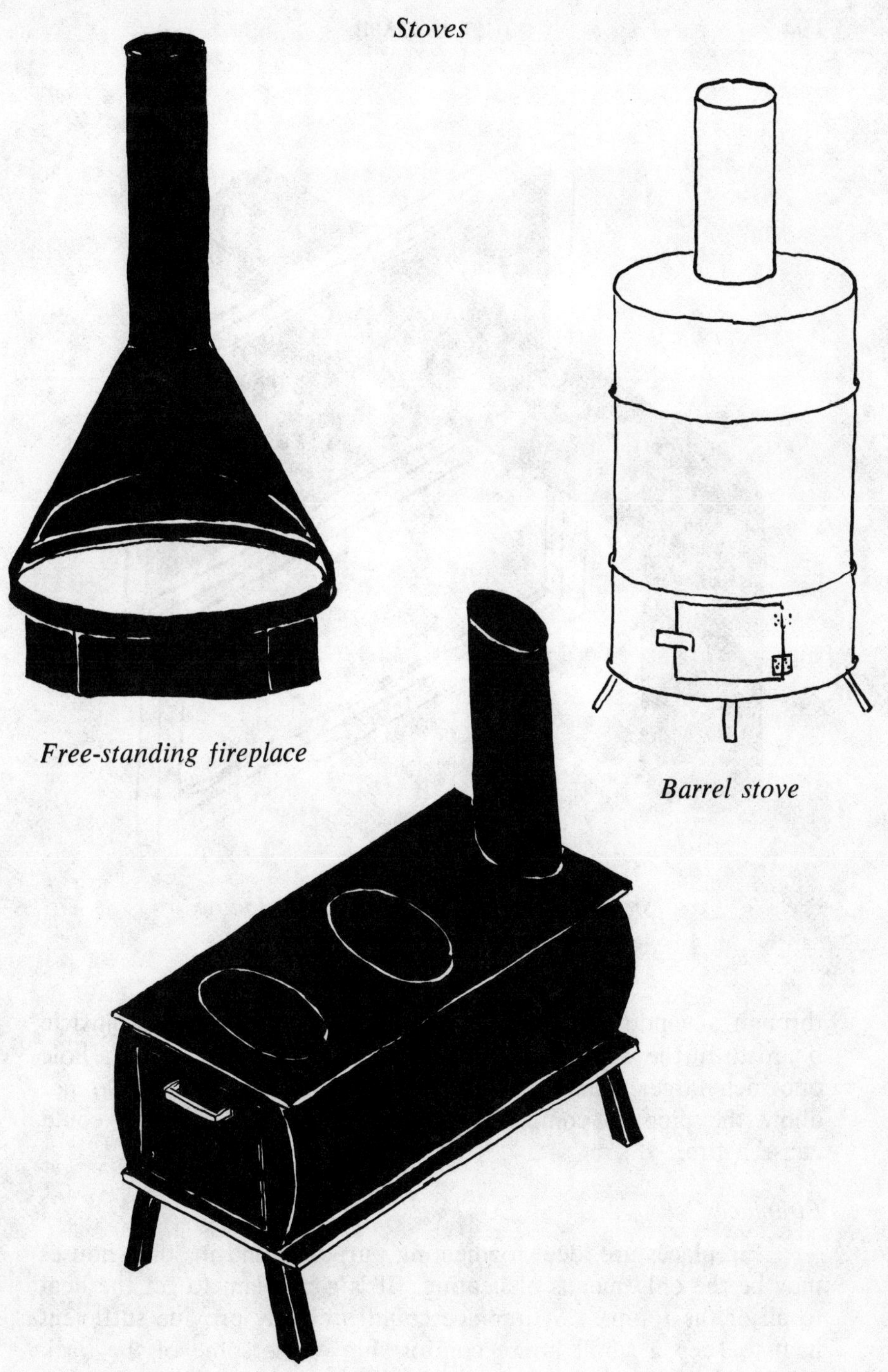

Free-standing fireplace

Barrel stove

Iron stove

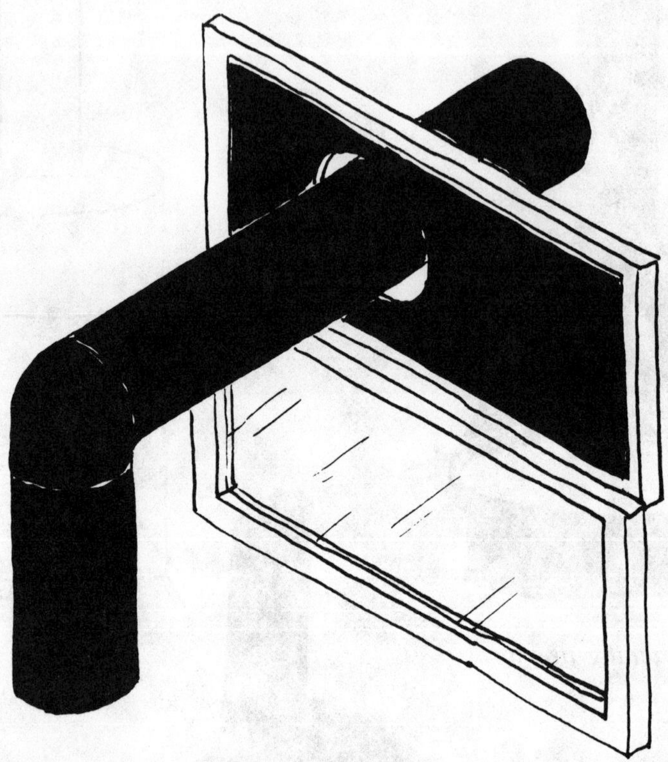

Stove exhaust through existing window

through a window, cut a piece of plywood, masonite or particle board to fit the window opening. Through this material cut a hole one inch larger than the diameter of the exhaust pipe. Do not allow the pipe to come in contact with the wood as it could cause a fire.

Fireplaces

Fireplaces are ideal for heating purposes, and in some houses may be the only means of heating. It is a problem to get the heat to all of the rooms. A fireplace could probably provide sufficient heat to keep a small house comfortable, while some of the back rooms might remain cool in a larger house.

No need to be concerned if your house was not built with a fireplace. Many different styles and sizes of free-standing fireplaces are on the market. You can find one to suit your needs. If you plan to make it a permanent fixture, use triple-wall pipe to go through the ceiling and roof. This could prevent a very serious fire.

27
Emergency Cooking Stoves

An emergency can be a frightening experience for anyone who is unprepared, but with a little forethought, you need never be caught short during an emergency. Learn how to take advantage of what you have: charcoal grills, cans, wire, barrels, aluminum foil, fireplaces, campstoves, canned heat stoves and heat tab stoves. There are many others which you can buy or create yourself.

As you gain experience in using emergency equipment, your confidence will grow, and with confidence comes security. This chapter will help you gain the knowledge that you need in order to prepare nourishing meals without the aid of modern conveniences and appliances.

Aside from possible availability of the commercially made charcoal grill, many different stoves can be made at home with materials that are on hand. Not much cost is involved and they work very well. The ideas presented below are not all original, but have been developed to meet the needs and requirements of my family. Look around and see what materials you have that you could transform into a handy emergency stove. Don't be afraid to experiment. The real test is cooking a meal on your invention.

Charcoal Grill

Many families have a charcoal grill. Whether it is an elaborate built-in unit or a small portable one, you can cook a full-course meal on it with little trouble. If you practice now, you'll be able to cook more easily when you're under the pressure of an emergency.

It would be wise to store charcoal briquets. The preceding chapter outlines a method for making your own charcoal.

Charcoal Stove

A charcoal stove is one of the easiest stoves to make. The materials needed are: One three-pound coffee can, three wire clotheshangers, six feet of soft (baling) wire, a can opener for removing lids of cans, and a can opener with a pointed end.

Remove the lid from one end of the coffee can. At the opposite end, using the can opener, punch holes in the side of the can. Completely encircle the can with holes, but do not connect them. The stove is now ready for use.

To make the grate for a charcoal stove, straighten out three wire hangers. Cut two of them into seven-inch-long pieces. Take the third wire and bend it in halves to make the handle for your grate. Now, weave the seven-inch pieces of wire and handle together in a waffle weave. With small pieces of soft wire, wire the grate together at each place where one wire crosses another. Twist the holding wires on the bottom side of the grate and cut off excess wire, leaving about ⅜ inch of twisted wire hanging straight down to keep the grate from slipping off the stove. To complete, bend the free ends of the grate wires toward one another as shown in the diagram. A little effort and about one hour of time. That's all!

Put in alternate layers of wadded-up newspaper and pieces of charcoal until the can is full, then insert a lighted match through one of the air holes at the bottom of the stove. If the paper burns but the charcoal does not light, empty the stove and repeat the process, using more paper.

Tin Can Stove

The buddy stove or tin can stove is generally used only for cooking, but could provide some heat for an hour or two. A one-gallon (three-pound coffee) can is all you need for the stove. Cut out one end of the can and slide it down against the other end. This doubles the thickness and helps to give more even heat. To hold the cut-out piece in place, punch a series of evenly spaced holes around the can. Toward the bottom of the can cut a side opening three inches high and four inches wide. This opening is used to insert the fuel cannister and also gives needed ventilation. Cooking is done directly on the can.

To make an oven to fit your tin can stove, use a shortening can. Remove both ends and wire a piece of see-through plastic roasting wrap over one end so that you can watch the food bake.

Emergency Cooking Stoves

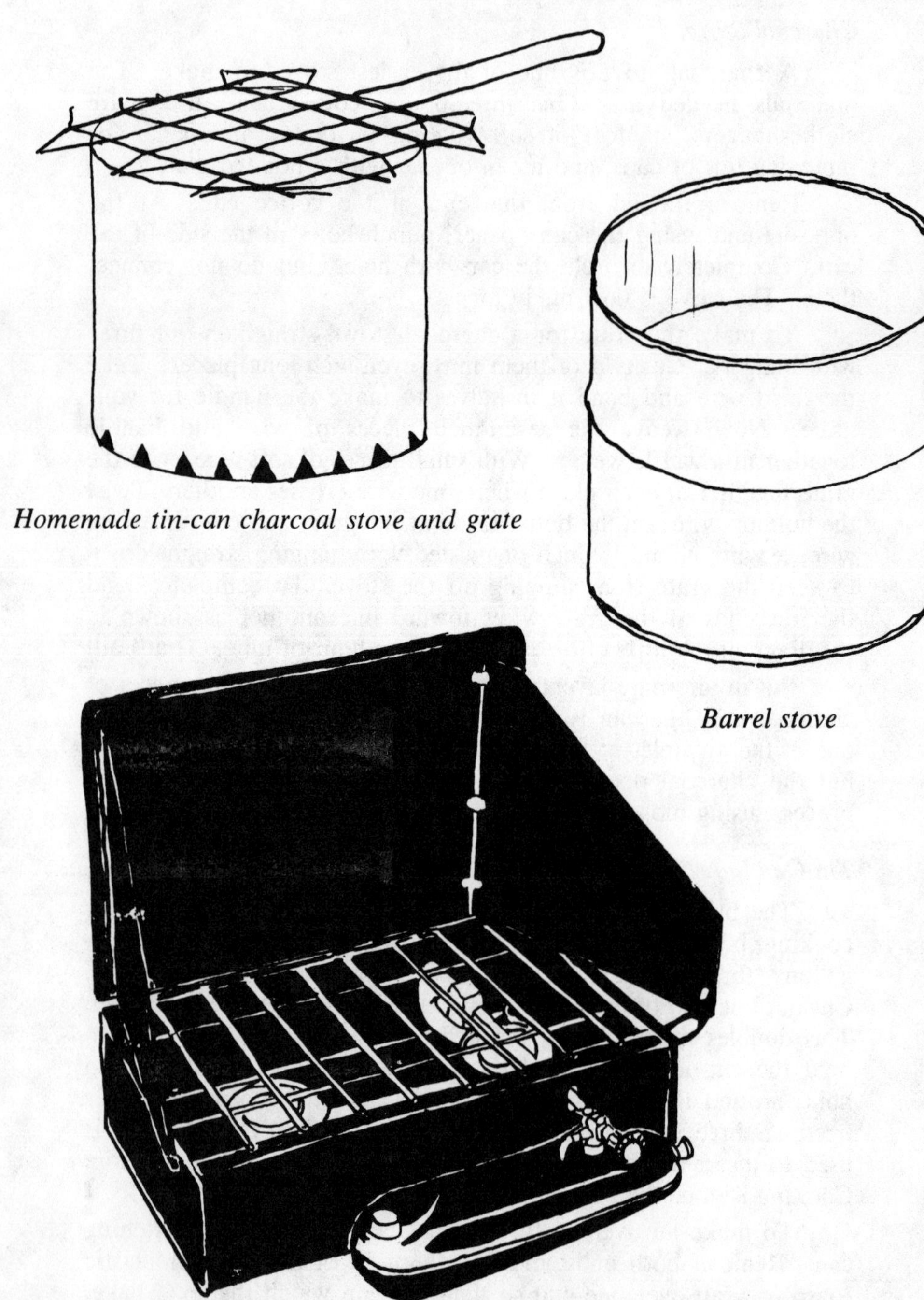

Homemade tin-can charcoal stove and grate

Barrel stove

Camping stove

Place the oven (plastic side up) on your stove. The heat will rise into it and cook whatever is placed inside the oven.

Fuel for Tin Can Stoves

Fuel for tin can stoves is easily made at home. A tuna fish can makes an ideal container, and the lid can be used as a damper. Cut a strip of cardboard as wide as the can is deep. Coil the cardboard and place it in the tuna can. Continue this process until the can is filled, then melt a small piece of paraffin wax and pour this over the cardboard. When the wax is cool and solidified, the fuel is ready to use.

If no other fuel is available, a small candle can be used for heat in the tin can stove.

A damper is used to control the flame produced by the fuel cannister. To make a damper, use the removed lid of the tuna can. Straighten out a wire hanger's bends and kinks. Punch two holes in the lid, parallel to each other and just large enough for the cut ends of the coat hanger to pass through. Cut a length of coat hanger twenty inches long, double the wire in halves, then bend down an inch and a half of the doubled end and half an inch of the two straight ends. Insert the two straight ends in the two holes previously punched in the lid, then turn the lid over and bend the protruding ends of the coat hanger flat against the lid. Punch two more holes on each side of the hanger and, with a fine wire, wire the handle to the lid as shown in the diagram.

Barrel Stove

Using whatever size barrel you can find, cut it in halves lengthwise. This is your stove. Make a grate out of quarter-inch metal rods, wired together or, if you have access to a welder, welded together. Place the rods one inch apart in each direction, with the vertical rods on top of the horizontal ones. A small grill could be made of quarter-inch metal plate and laid at one end. Wood, coal, charcoal, paper or just about any other available dry fuel will burn in this stove. If you prefer, this stove can be made by cutting the barrel in halves horizontally.

Aluminum Foil Cooking

Perhaps cooking in foil is the easiest way of all. You need only a piece of aluminum foil (preferably heavy duty) and a fire. In the center of a piece of foil about fourteen inches long, place a dab of butter. Slice vegetables about one-eighth inch thick, then arrange a layer each of carrots, potatoes, onions, and any type of meat you desire. Again make layers of onions, potatoes and carrots. Add another small dab of butter to the top of the stack. Other vegetables can be used, according to your taste. Place the vegetables that take the longest to cook next to the foil, with the meat always in the center. With this arrangement, the juices will flow to all the vegetables during the cooking period. Add seasonings to your taste, then fold the ends of the foil to the center, making a seal. To help prevent puncturing and losing the delicious juices when you turn the package in the fire, wrap another piece of foil around the package, sealing it in the same way as the center package.

Place the foil package in the fire and cover with coals. Allow it to cook for about twenty minutes, then turn it over and cook it for another fifteen minutes. Remove from the fire, unfold the foil and open the package, rolling the edges as you go. Be careful of the hot steam that will escape, and don't let the juices spill. Eat right from your aluminum baking dish.

Aluminum pans from frozen dinners can be used for cooking dinners if they are sealed well.

Dutch Oven

For a one-pot meal, you can't beat a dutch oven. Simply place the food in the dutch oven, set the lid on, and put the whole thing in the fire. Cover it with coals and keep a good fire going around it. Occasionally, check to see if the food is done.

You can bake, fry, boil or broil in a dutch oven, so long as you have a fire with coals.

Spit Cooking

One way of cooking delicious food is with a spit or stick. Pierce the food with the spit and place over the fire. Rotate the spit or stick slowly until food is cooked, then remove and eat. If a stick is used, select a green, straight one, and remove any small twigs or leaves.

Camp Stoves

Camp stoves are handy in an emergency. They come in one-, two- and three-burner styles. Probably the most common in use today is the two-burner stove. Once set up, a camp stove is as easy to use as a kitchen stove. Those using white gas as fuel have a fuel tank that is pressurized by a built-in air pump. After pressure is brought up in the fuel tank, the stove will burn with an even, constant flame. Occasionally the fuel tank will require repressurizing (pumping). Almost all camp stoves have knobs for adjusting the height of the flame.

Another type of camp stove uses propane fuel. This stove takes a little less work as there is no need to pressurize the fuel tank with a pump. Just connect the fuel hose to the burners and it is ready to use.

Heat Tab Stove

The heat tab stove is a very small, three-legged stove with a cup in the center just large enough to hold a one-inch diameter by half-inch thick fuel tablet. Many backpacking campers use this little stove on overnight hikes. It folds into a compact unit that fits into a very small space. While it is generally considered to be a one-man stove, it could serve in a short-term crisis to heat food or warm a baby bottle.

Canned Heat

Canned heat is a fuel that comes in a small can, and when lit provides sufficient heat to cook a meal. When the cooking is finished, you just slip the lid in place and the flame is snuffed out. If the lid is kept tightly closed, the fuel will store for many years. I have a few cans that were kept for fourteen years. Upon opening one of the cans, I found that about a third of the fuel had evaporated. However, time had not diminished the strength of the remaining fuel. When a match was placed in the can, it ignited immediately. Most manufacturers of canned heat also market a small collapsible, one- or two-burner stove to use with the fuel.

Other Equipment

There are a few stoves on the market that would be nice to have in an emergency. Some are the sheepherder, Kwick-Kamp and Sims stoves. These come with or without legs, ovens, and shelves which extend the width of the stove top. Check with your local sporting goods store for details on these and other stoves.

Open Fire

The oldest fire used for cooking or heat is the open fire. To make a fire ring, select rocks or bricks (fire brick, if possible)

Commercial reflector oven

Open fire

Dutch oven

Homemade cardboard-box reflector oven

and place them in a circle. Construct the fire in the center and build a good set of coals. Place your pan on the coals and cook in the same way as you would with any other fire.

Caution: Do not place rocks *in* the fire. Rocks have moisture in them and when they get hot they will expand and explode, throwing rock fragments in all directions. Before building a fire ring, make sure you are not under a tree or bush. Clear the immediate area of any brush, paper or other flammable material.

Reflector Oven

When using a fireplace or an open fire for cooking, and you want to bake something, a reflector oven may be used. This is a small heat shield that reflects heat from a fire. Reflector ovens are available wherever camping equipment is sold, or you can make your own from aluminum foil, a cardboard box, and a few wire coat-hangers. Lay a cardboard apple box on its side. Cut a flat piece of cardboard the inside width of the box and three inches longer than the height. This piece will be the reflector backing. Cover the inside of the box and the reflector with foil, then bend the reflector lengthwise in the middle to form a V and insert this in the box with the base of the V pressed tightly against the bottom of the box and the ends of the V wedged against each end. To make shelf supports, straighten five wire hangers and push them through one side of the oven, at a height parallel to the V of the reflector, and out through the opposite side. Let about an inch of the coat hangers stick out on each side of the oven. These ends can be bent down to prevent them from slipping out of the holes. Last, cut a flat piece of cardboard the width and depth of the oven, cover it with foil, and rest it on the wire shelf supports.

Stand the oven in front of the fire and heat from the fire will be reflected so that the food is cooked on both sides.

With the few ideas in this chapter and a little preparation, you should be ready to serve a ten-course meal in the middle of a . . . Well, at least they may help you to be prepared enough that you can boil some water and cook a good soup.

28
If the Light Switch Fails

The prospect of nights without light is not a happy one. One night might prove to be an enjoyable adventure, but most of us would find it difficult to care for ourselves and our families for very long in the dark. How attached to the convenience of light we have become!

For the future, it appears that we can look forward to more summer power shortages because of increased demand for electricity for cooling. In the winter, when furnace thermostats are set high, there is the added risk of severe storms which may cause power outages. In these circumstances it would be wise to provide some manner of lighting the home in an emergency. In addition, if you are aware that you will have no lights for several nights in a row, you could plan to do in daylight such things as laying out night clothes, bathing, etc.

Candles

Candles? Are they practical in today's world? Perhaps the slim, formal ones are not, but good, solid, functional ones happen to be extremely practical. In my particular neighborhood, for example, when the wind becomes stronger than a gentle breeze, I automatically set out the candles and matches on the table and the counter. And when the wind sounds a certain way, I *light* the candles — before the lights go out! Quite often the power does go off temporarily, so candles have become part of our way of life.

Eating by candlelight by choice is fun. Wishing you had a candle for light in an emergency is another matter.

Creating candles at home. A variety of candles can be made

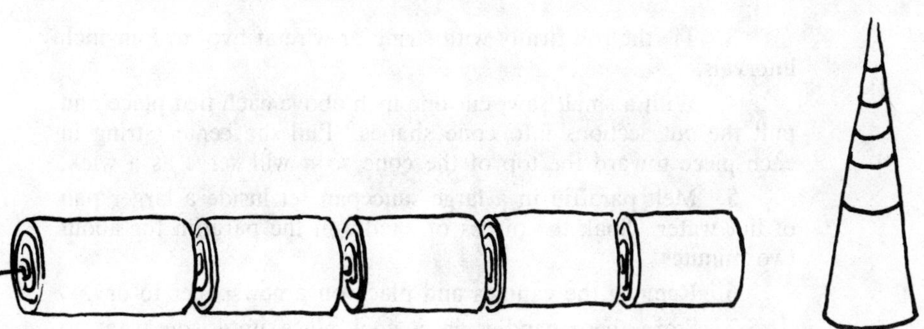

Trench candles

fairly inexpensively at home. They range from strictly utility candles to the kind you could put on an elegantly set table.

Perhaps the value of the homemade candle is greater than we realize. This delightful comment about candles was written in the early twentieth century.

> Professor Thompson, a celebrated electrician, declares that if the electric light had existed for centuries and the candle was newly invented, it would be hailed as one of the greatest discoveries of the age, being entirely self-contained, cheap, and portable, and requiring no accessories in the way of chimneys or shades . . . For all these reasons, as also on account of the usefulness of a bit of candle for the homelier purposes of the household, it is worth while to revive the art of candlemaking. Time often hangs heavy on the hands of the boys and girls, especially on stormy days in the country, and there can be no harm in turning a quantity of mutton or beef suet and a little beeswax into a dozen or two candles, which will save kerosene and which would cost quite a little sum at the grocer's.[1]

Trench candles. One of the easiest and most inexpensive versions of an emergency candle, the trench candle also can be used as emergency fuel. It is made from rolls of newspaper soaked in paraffin.

1. Place a narrow strip of cloth or twisted string (for a wick) on the edge of newspapers of six to ten layers.
2. Roll the paper very tightly, leaving about three-quarters of an inch of wick extending at each end.

[1]Sidney Morse, *Household Discoveries, An Encyclopedia of Practical Recipes and Processes* (New York: The Success Co., 1908), p. 106.

3. Tie the roll firmly with string or wire at two- to four-inch intervals.

4. With a small saw, cut one inch above each tied place and pull the cut sections into cone shapes. Pull the center string in each piece toward the top of the cone so it will serve as a wick.

5. Melt paraffin in a large saucepan set inside a larger pan of hot water. Soak the pieces of candle in the paraffin for about two minutes.

6. Remove the candles and place on a newspaper to dry.

7. Store these candles in a cool place until you want to use them.[2]

Wax candles. Information given in this section on candles will be for a basic utility candle. The hobby and craft shops have instructions available for making the fancy ones.

Only two items are necessary for making candles — suitable wax and a proper wick. Your equipment need not be expensive or fancy, and even the pouring process is quite simple. So why not make several functional candles to have ready — just in case.

Wax. Several kinds of wax are used in candlemaking, the difference being in their hardness. Wax comes in eleven-pound slabs, in chunks, and in one- to two-pound slabs. The cost of wax may be less if it is purchased in a hardware or discount store rather than in a craft shop.

Grocery stores usually have a supply of household wax with the canning and freezing supplies. This inexpensive wax is sold in one-pound boxes containing four individual bars of wax. It takes about two and a half pounds of wax to make a candle seven inches tall and three inches thick.

When using a large block of wax, break it into usable pieces by putting the block into a large, heavy grocery sack, placing something under one end to raise it two or three inches off the level surface, and giving it a couple of good whacks with the *side* of a hammer.

Wicks. You can buy wicking or you can make your own wicks. Cotton or braided wicking is available wherever candle supplies are sold. Wire case wicks are also available. Hardware stores have carpenters' or bricklayers' chalk line, which makes a good wick. You can use wicks from broken candles; or, if you

---

[2]*Girl Scout Handbook* (New York: Girl Scouts of America, 1953), p. 271.

are making large, sturdy candles, use cheap utility candles for the core.

To make your own wicks, use medium-weight cotton string, and soak it overnight in a solution of two tablespoons of Borax plus one tablespoon of salt dissolved in one cup of water.

One way to insert a wick is to thread it through a hole made in the finished candle with an ice pick or hot wire after it has cooled. After you have inserted the wick, pour a little more melted wax in around it to make sure it is held firmly in.

Another way to put in a wick is by placing a pencil or dowel across the top of the mold. Tie the wick securely around the dowel and let it hang down into the center of the mold. After the candle has cooled, simply cut the wick off close to the dowel.

Molds or forms. Any suitable form can be used, such as empty milk cartons of all sizes, empty food cans, glass jars, heavy paper, plastic containers. Just about anything that will hold the wax can serve this purpose.

If you plan to use food or juice cans, select those that are smooth and straight inside. Any ripples or ridges will prevent the candle from sliding out after the wax is hard.

If glass jars are used, they must be broken away after the wax has cooled. Empty pickle or jelly jars serve well as candle molds. After the wax has cooled and hardened, place the jar in a heavy grocery sack and whack it soundly with the side of a hammer a couple of times. You don't want to pulverize the glass, just break it up. After the glass is broken, reach in *carefully* and remove the candle, protecting your hand with a pot holder or paper towels. There is usually no problem unless the glass is broken too much. It generally falls away from the candle in large chunks. Of course, taking the candle out of the bag is *not* a job for a young child!

I sacrificed several "real" canning jars at Christmas-time last year and made candles for the grandparents this way. One grandma thanked me for "homemade jam" before she realized what it was.

By adding a few crayon pieces to color the wax, these candles can be attractive as well as functional.

Getting candles out of molds. Paper molds can be carefully torn off. Wax comes out easily from a plastic mold.

Mold release, purchased in hobby and craft shops, is needed

on metal molds; or ordinary cooking oil can be used, though it sometimes causes mottling or blotches on the hardened wax. If the candles are for utility purposes only, a few spots won't hurt anything.

Equipment.

Wax

Mold

Oil or mold release (for metal molds)

Wicking

Double boiler (or a pan with a lid and larger pan to hold water and serve as a double boiler)

Pot holders

Wooden spoon

Ice pick

Large tub, bucket or wastebasket — for cooling candles

Dowels or pencils

Basic instructions.

1. Prepare molds. If using a metal mold, make sure it's clean. Coat the inside with cooking oil or mold release, and blot off excess oil.

2. String the wick (unless you plan to use an ice pick after the wax has cooled). Weight the string so that it will not float. Small fish sinkers work well as weights. Tie the wick around a dowel and place it across the top of the can, the weight touching the bottom. The wick should be straight, not slack or curled.

3. Melt the wax, either in a double boiler or in a small pan placed in a larger pan of water. Always keep water in the bottom pan.

4. Heat the wax slowly, until all of the wax is completely melted. *Do not leave melting wax unattended!*

5. When max is melted, pour slowly and evenly into mold, reserving some melted wax for use later. Fill the mold to within two to three inches of the top.

6. In order to cool the wax evenly, place the mold in a bucket or tub of cold water. Put a weight on the mold to keep it from tipping over. The water level should be two inches higher than the wax level.

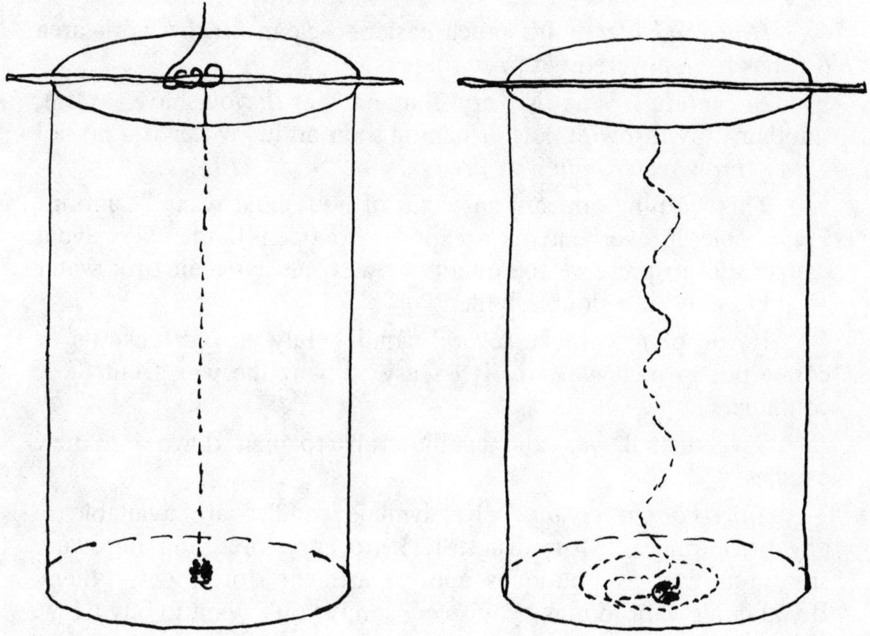

Placing wicks for candles

7. Allow the candle to set up six to eight hours in the water, then remove the candle from the mold.

Candles without molds. It is extremely easy to make candles without molds. They require a one-pound box of household wax, wicking, and a small amount of melted wax to work with.

1. Fuse together two quarter-pound blocks of wax by dipping a side of each in melted wax, then pressing together.

2. With a sharp knife or ice pick, scratch a groove lengthwise in the center of one side of the fused cake, and insert the wick.

3. Dip the other two quarter-pound blocks of wax in the melted wax, and add them to the other fused block.

4. Fuse the ends as well as the sides by dipping them in the melted wax.

5. The sides should be built up with the melted wax, as it cools, to form a well around the wick.

6. This kind of candle can be built higher by adding more blocks and stacking them in the same manner.

Hints and ideas. It's much easier to clean up the work area if it has been covered with newspapers.

Be careful! Wax is very flammable. If you have a fire, smother it by throwing lots of baking soda on it. Water is a no-no! *Never throw water on a wax fire.*

This warning appears on boxes of household wax: "Caution: Flammable if overheated or exposed to open flame. To avoid danger and to preserve the quality of wax, always melt over water — preferably in a double boiler."

If you plan to make several candles, buy an old teakettle or coffee pot to melt wax in. It is easy to pour the wax from these containers.

Save melted wax and candle stubs to melt down into new candles.

Store-bought candles. Readymade candles are available in any price range. Most drugstores, grocery stores and hardware stores, as well as boutiques and department stores, carry them. If you don't want to make your own, surely you'll want to buy a few.

The burning time will vary, depending on the size of the candle, the type of wax, and the size and kind of wick.

Approximate burning times are as follows:

¾ inch diameter, 4 inches tall — will burn 2½ hours.

⅞ inch diameter, 4 inches tall — will burn 5 hours.

2 inches square, 9 inches tall — will burn 7 hours per inch.

2 inches square, 4 inches tall — will burn 28 hours.

2 inches square, 9 inches tall — will burn 63 hours.[3]

Though candlelight is romantic it does not give much light. With this in mind, you may need more than one candle for that "some day." If you start now, you can buy or make enough for a good supply.

Candle holders. Anchor candles firmly to a broad-based object to prevent accidents. Some candleholders have a small spike protruding upward to hold a candle firmly in place. A cup, saucer, glass or jar lid will serve equally well if you allow sufficient melted wax to drip onto the bottom of the holder so that the candle will stand firmly when it is placed in the wax.

---

[3]Zabriskie, *Family Storage Plan,* p. 51.

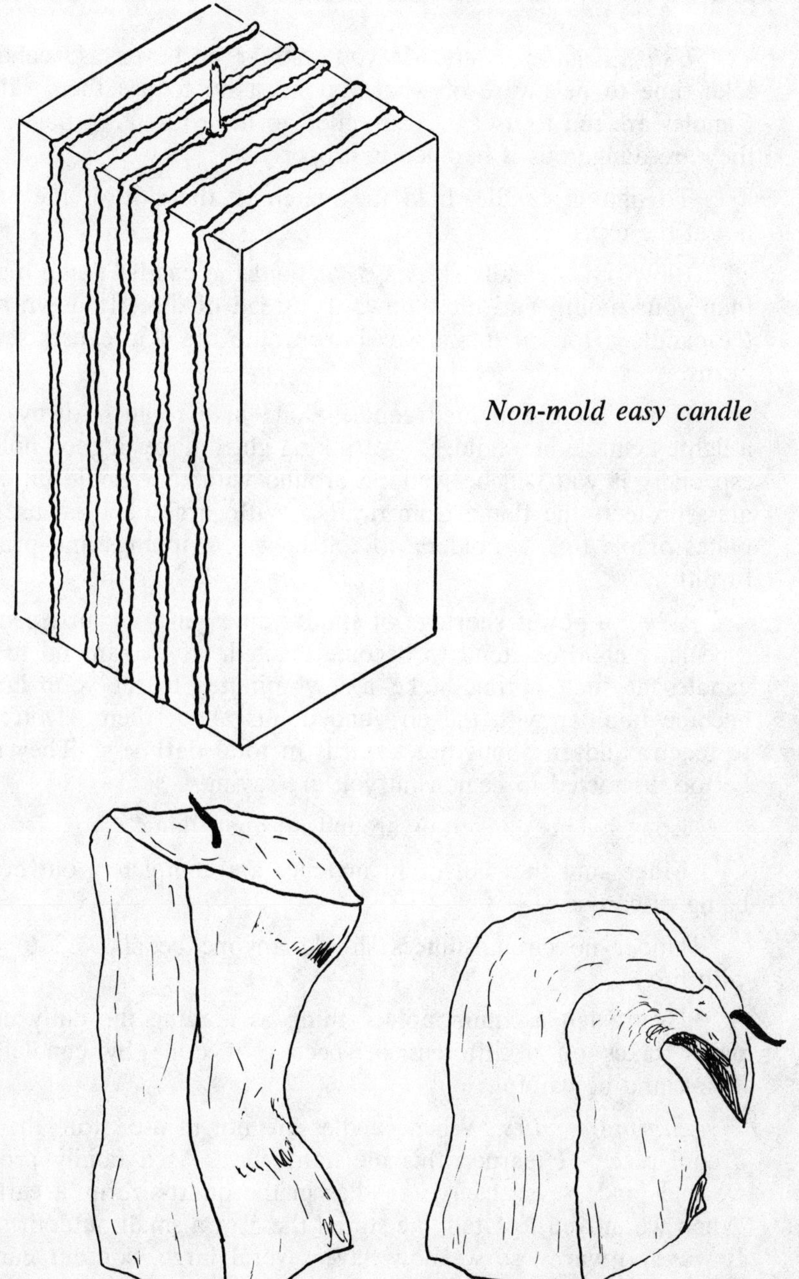

Non-mold easy candle

Improperly stored candles

Using candles safely. If you seldom or never use candles, take time to be aware of what you must do to use them safely. Candles are fun to use; they are good to have in emergencies; but they are dangerous if handled improperly.

To light a candle, hold the match to the side of the wick, not at the top.

Blow out a candle sideways, or hold the candle flame higher than your mouth and blow upward instead of directly down onto the candle. Hot splattered wax blown onto the skin causes severe burns.

Never leave a lighted candle alone in a room. Always put a lighted candle in a holder. A drinking glass makes a good holder, especially if you will be walking around with the candle lit. The glass protects the flame from drafts. With large candles, use old plates or pie tins as holders, to collect wax drippings and protect furniture.

When a power shortage or similar emergency occurs, people, especially children, tend to become excited. If you intend to use candles at such a time, take a few minutes to let your family become familiar with the do's and don'ts about them. Don't try to teach children about fire hazards in total darkness. They may be too distracted to hear what you are saying.

Long hair is dangerous around an open flame.

Make sure that hot or lit matches are completely out before being discarded.

Under no circumstances should anyone be allowed to play with fire.

Even such a commonplace thing as reading the daily newspaper takes on a different perspective if done by candlelight. Think and be careful.

Storing candles. When candles are not in use, store them in a cool place. I learned this the hard way. As a family project, we had made several large candles in the quart-size milk cartons. When we moved to Utah, we stored these in a small outdoor shed. It was too warm, so we now have several large, contour candles that look strange, but still burn.

When storing several candles together, use paper between each layer.

Other uses. In an emergency, candles come in handy for things other than light. Small candles are good for lighting outdoor fires, because the flame remains constant until the wood catches fire. They can be used for cooking with a tin can stove. Trench candles can be burned as mini-logs for fuel. (See preceding chapter for details on these last two suggestions.)

Kerosene Lamps

The sight and smell of kerosene lamps conjures up memories of my grandparents' farm. I learned that you didn't just wash a lamp chimney; you had to shine it with newspaper. Grandpa and Grandma had gorgeous lamps, all flowers and frills, with colored opaque glass. I learned too, that the lamps are hot, and that they have a smell all of their own.

Availability. Kerosene lamps are available today in most hardware stores. Some grocery stores carry them, as do many army-navy surplus outlets. Promotional sales for lamps begin when camping equipment is on sale. Utility-type kerosene lamps are not expensive, cost varying with the size of the lamp.

Light from a kerosene lamp is far brighter and steadier than that from a candle, and it can be adjusted to varying degrees of brightness. A lamp is sturdier and more stable than a candle.

Care and use of lamps. If a lamp is smelly or gives poor light, it is because it is not kept clean; the wick is clogged from having been used too long; or the chimney is wrong.

If a lamp is used often, trim, clean and fill it daily, then wipe the whole lamp.

If you trim the wick by rubbing the char from it, this leaves it even. You can't *cut* it even.

Clean the burner out quite often with a good solvent, keeping the holes in the floor of the burner clean for air.

Do not completely fill the lamp, since the oil expands with heat and will run over. Empty the font occasionally to clean out the sediment.

Do not open a lamp when it is hot. There is explosive vapor in it.

Light the lamp with the wick turned low and turn it up gradually, so that you do not get it too high and make smoke.

If a lamp has been burning long enough to get hot, move it with care. Better still, don't move it at all.

One common cause of explosions is the upsetting of a lamp. To avoid this problem, choose a lamp with a broad, solid base, one that is not top-heavy.

Never fill a lamp while it is burning. This cannot happen if you buy lamps that have no openings except the one for the wick.

Do not blow out the flame without turning down the wick, especially if the lamp has been lit for some time. With most lamps, when the wick is turned down far enough the flame will go out.

Always turn the lamp down low when carrying it about.

Chimney.

> The object of the lamp chimney is to supply the flame with exactly the amount of air it needs for perfect combustion, no more and no less, with an even draft on both sides of the flame. They must be clear and transparent. This calls for fit in the full meaning of the word and for clean glass that will stay clean. Thus there is something to know about chimneys beyond the mere size of the bottom. The ordinary notion of fit is a chimney that will stay on the lamp and not fall off. That is part of the fit. The rest is such a shape as to make the right draft for that particular lamp. It includes the seat, bulb, shaft, proportion, sizes in all parts and length. Good chimneys that fit will give more light than common ones.[4]

When replacing a broken chimney, use the one recommended by the manufacturer if possible. If you jot down and keep the measurements before your chimney *breaks,* it will be easier to get the right replacement.

Kerosene as a fuel.

Safety note: Do not store kerosene or other such fuels in the home or where children can reach them! Store kerosene in a tightly closed metal container, in a very cold place. It is highly combustible.

Kerosene is relatively inexpensive as a lamp fuel and burns easily. It is available at some gas stations, sporting goods stores, and fuel depots.

Scented oil as a fuel. The colored and scented oil that can be purchased as fuel for these lamps is much more expensive than regular kerosene, but storage rules are the same — *do not* store it

---

[4]Morse, *Household Discoveries,* p. 102.

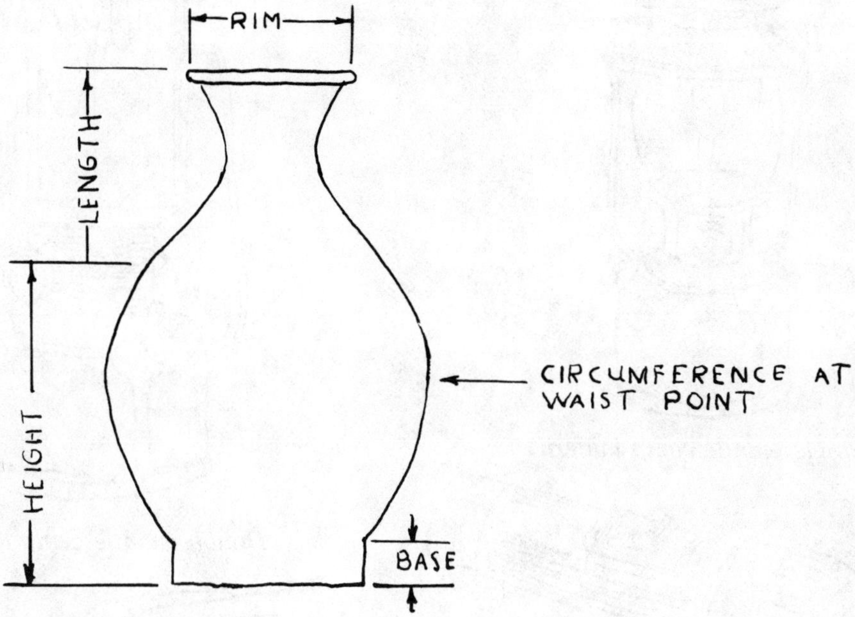

How to measure chimney

in the house. Because it looks and smells pretty, it may be an attraction to children, so keep it out of their reach!

Wicks. Replacement wicks can be purchased at most hardware stores and usually at army surplus outlets. Wicks become clogged and should be replaced if they have been standing in oil or used for several months, because they don't then feed the kerosene freely.

Camping Lanterns

Another light source, which you already may have tucked away with your tent and sleeping bags, is a camping lantern. It provides a fairly bright light, and can be purchased with a single or double mantle, the mantle acting the same as a wick. A lamp with a single mantle is smaller and gives less light than one with a double mantle. Camping lanterns burn white gas. They have been constructed for outdoor use and generally are sturdy and well balanced, but the safety precautions given for a kerosene lamp also apply to a camping lantern.

Commercial Lights Available

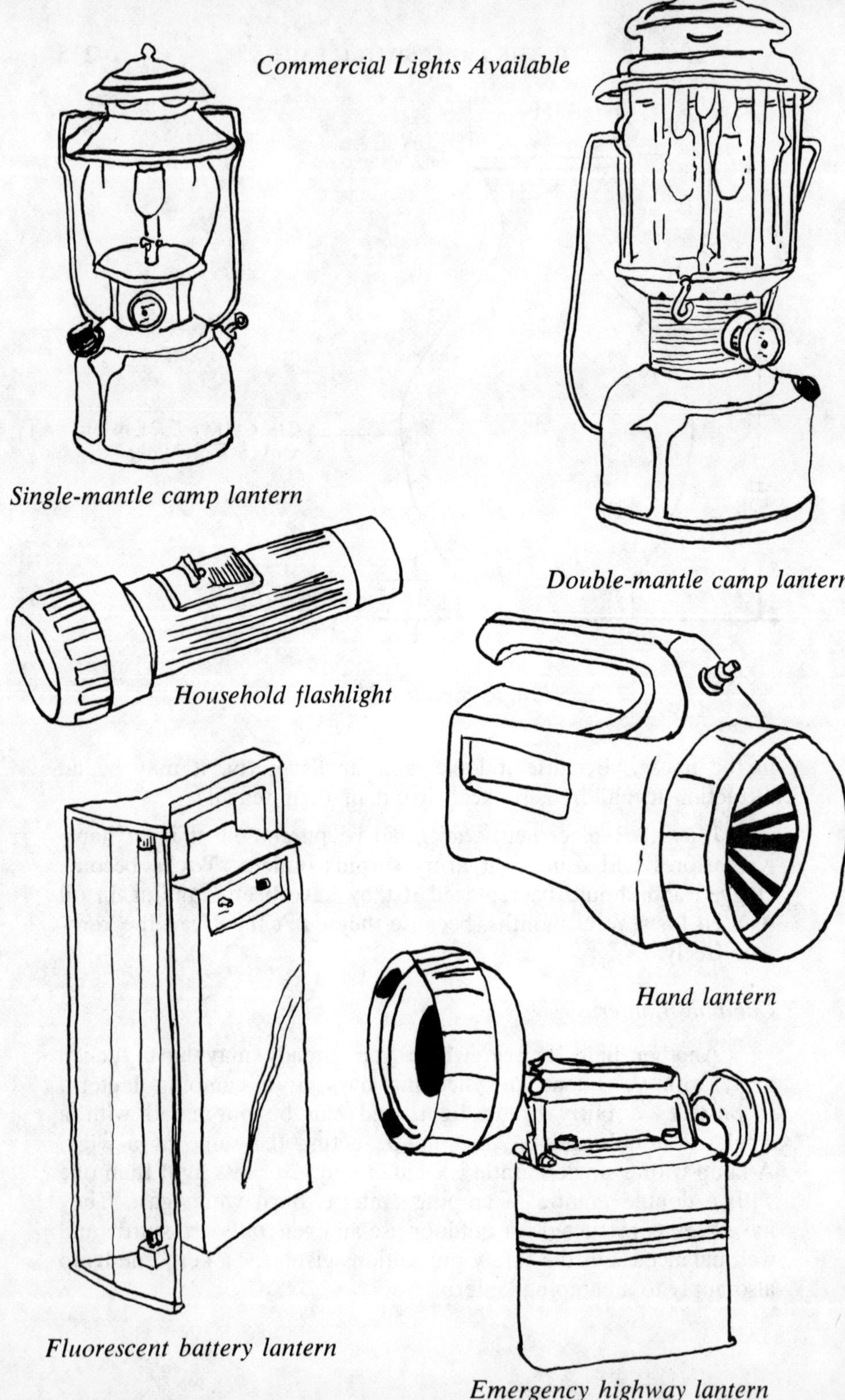

Single-mantle camp lantern

Double-mantle camp lantern

Household flashlight

Hand lantern

Fluorescent battery lantern

Emergency highway lantern

Lanterns are now available that burn propane. Another type of lantern now being marketed uses fluorescent tubes and operates on batteries. A number of different types of lanterns available for use as part of your camping equipment would serve well in an emergency situation.

Availability. Lanterns, chimneys, mantles and fuel are available at most sporting goods stores, sporting goods departments or department stores, hardware stores, and usually at army surplus stores.

The cost naturally varies with the size and type of lantern.

White gas as a fuel. Follow all safety precautions when using white gas. Filling lanterns and lamps with flammable fuel should not be a child's chore. White gas is comparatively inexpensive as lantern fuel. It is available at sporting goods centers and some gas stations.

A Fire

Fire is unpleasant to think about, but it does happen occasionally with candles, lamps and lanterns. Be prepared. *Think it out beforehand.* What would you do if . . . ? If it happens, try not to panic. If a lamp has been tipped over and is burning, *don't* throw water on it. Grab a blanket or towel, wet it, and beat out the flames. Water would spread them. Close doors and windows quickly to cut off drafts. If just the lamp is burning, grab it and throw it outside. If it's a small fire, smother it with baking soda or wet rags.

Flashlights

Flashlights are another emergency source of light, but if they are to furnish your only light for a short time, consider how many individuals will need light, and where, and for how long. Perhaps you will need more than one flashlight.

Flashlights come in any number of sizes and vary greatly in light strength. Their cost varies, of course, with their size, style, and how sophisticated they are. All flashlight equipment is available in virtually all drugstores, discount stores, and hardware stores.

Battery and bulb storage. Batteries can be stored for as long as eighteen to twenty-four months if kept cool and dry. They retain their strength much longer if left out of the flashlight and kept from touching one another.

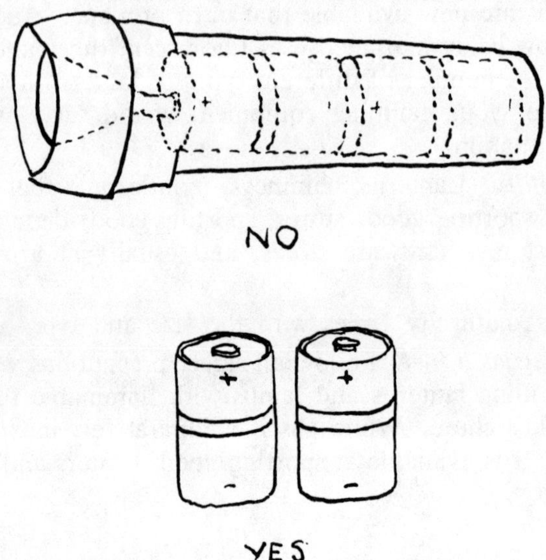

Proper storage of flashlight batteries

Bulbs last longer if they are stored where they will not be bumped or jiggled, causing the tiny filaments to break.

Plan Ahead

Though it might be brighter, a flashlight will not last as long as candles or lamps. With continual use, a flashlight, running on new batteries will burn for only about six or seven hours.

It might be best to be able to combine one or more lighting methods. For a short time, perhaps a candle or flashlight would do; for an emergency of longer duration, you might prefer a lamp or lantern.

Around our house, when that ol' wind blows and "things go bump in the night," it's certainly more cozy with a light!

29
Soap Making

Many people think that making soap at home is part of past history; that it has been a thing of the past since pioneer times. Well, if you think that way, you're in for a surprise! People still make soap at home and you can too.

Soap is actually very easy and inexpensive to make. Recipe ingredients vary and the results range from perfumed hand soap to laundry soap. In time of need or emergency it would be well to know how to make a variety of soaps. Besides, it is fun to say, "I know how to make soap."

Essentials of Soap Making

> Essentially soap is made of two ingredients — fat and lye. These two ingredients are mixed together at a temperature most favorable for starting the soap making reaction, which is called saponification. It is a chemical reaction — the ingredients come in contact with one another, mingle, interact and turn into a new product that is a combination of soap and glycerine. Though the glycerine is often removed from manufactured soaps, homemade soap retains its glycerine, making it all the nicer.
>
> The reaction begins in the pot in which the fat and a solution of lye and water are combined. The mixture is stirred continuously while it thickens, and the soap begins to form. The soap is not left to harden in the pot but poured into molds. Here the saponification continues and — eventually — ends.[1]

Common Ingredients

Soap can be made with just fat and lye, but it is usually a combination of several ingredients, descriptions for which follow.

---

[1]Ann Sela Bramson, "Make Your Own Beauty Soaps," *Family Circle* (New York: The Family Circle, Inc., 1974), p. 76.

Lye is a white caustic solid that is available at most supermarkets and hardware stores in granule or crystal form. *Caution:* Be extremely careful when using lye. It causes severe burns very quickly if it gets on the skin. If at all possible, never work with it when children are present.

Borax, a white crystalline salt used to promote the blending of products, can be bought at the drugstore.

Glycerine is made from fats and oils and is produced when making soap. It is used in the manufacture of some commercial soaps. Glycerine added to homemade soap gives it a more luxurious feel.

Fats and oils (vegetable oil and most animal fats) are used to make soap.

To Render Fat

Vegetable oils and lard can be used as they are, but animal fat, grease or drippings must be rendered before they can be used for making soap. This means boiling the fat or drippings in water until pure grease rises to the top; any impure residue will sink to the bottom.

Put the animal fat, grease, or drippings in a large kettle and cover it with twice as much water as you have fat. Bring the water to a boil, and boil it for thirty to forty-five minutes, depending on the volume of fat to be rendered. As it cools, the clean grease will rise to the top and can be skimmed off.

Please note: Home-rendered grease should be made into soap within twenty-four hours! The first time I tried making soap, it was from scratch, including rendering the fat. I had some large pieces of animal fat, which I boiled in a very large kettle. As expected, the grease rose to the top, and I skimmed it off and put it into a can to use "tomorrow." "Tomorrow" came and I was sidetracked by something else; the next day also. By the time I got around to using the grease, a crust had formed on top. Unconcerned about this, I deftly scooped it into the canner to start making soap. It had become rancid — very rancid! It had to be buried, then I opened all the windows and doors in the house, but I still had to leave for several hours while the smell faded. The lesson I learned was to use rendered grease immediately. My *next* batch of soap turned out quite nice — I cheated and bought lard!

Scents, Oils, and Perfumes

Aromatic oils and scents make hand soap more luxurious but are not essential ingredients. Even when you use these "extras," homemade soap remains inexpensive. Scents, oils, and perfumes are available in drugstores, potpourri shops, and some boutiques.

Equipment Needed

Most of the equipment needed for making soap is already in your kitchen. Special pans or spoons are not needed, since those that you use will wash out easily.

A large kettle, preferably enamel, iron or stainless steel. *Never use aluminum,* as lye reacts with aluminum.

A long-handled wooden spoon, for stirring the soap. A wooden spoon is recommended because the handle does not get too hot to hold.

Molds could include shallow wooden trays, glass pie plates or loaf pans, or a large cardboard suit box lined with a plastic garbage can liner. Do not use any aluminum utensils for molds.

A *scale* is used for weighing grease. It is nice to have one, but not necessary.

Soap-Making Hints

— Pour lye or lye mixtures as slowly and evenly as possible to prevent splashing and burning.
— A no. 10 can holds five to six pounds of clean grease.
— Ten cups of grease are equal to about five pounds.
— The soap is ready to pour when it reaches the consistency of thick gravy.
— When the mixture is ready to pour in molds, add scents, oils or perfumes, stir well, then pour.
— Beware of the fumes. Don't smell or sniff to "see how it's coming." Fumes are dangerous!
— Get your molds ready *before* you start mixing the soap.
— It takes about forty-eight hours for soap to set up. Mark and cut it into bars after the first day, but leave it in the mold until it is firmly set. If you do not cut the soap until it is solidly set up, it may not cut evenly. Other than making uneven bars, this has no effect on the soap. If you

remove the soap from the molds while it is still soft (but not runny) you can roll it into balls.

— Homemade soap should age for about two weeks before use. Allow air to circulate around it. "Aging gives the soap a chance to incorporate any unreacted lye that might be left in it."[2] Aging takes place after it has been cut into bars. Place it in a grocery bag in the closet.

— Soap can be grated after it has aged.

Economy

If you save and use your meat fat drippings, homemade soap is unbelievably inexpensive. The cost of my soap was only about a fourth as much when I used my own grease, as compared to buying the lard. One batch makes about twenty-two large bars of soap.

Some homemade soaps can be grated and used for dishes and laundry, thus becoming an all-purpose soap.

Here is an unsolicited commercial from a friend who uses her homemade hand soap for the laundry too. "This soap is really good and it does make our clothes clean. My neighbors don't like to hang their clothes out on the same day that I do because mine are so white and bright."

If you use soap to launder your clothes, remember that it is pure soap and must be rinsed out thoroughly.

Some Special Soap Recipes

Try making soap, even if it is just for fun. If you find that you like using it, it can surely help your budget. And you'll have the knowledge in reserve — just in case.

Five soap recipes are offered here. You might try several to find the ones that will suit your purposes.

Grandmother's Homemade Soap

12 cups of clean grease
19½ cups warm water
1 can lye
¼ cup ammonia
½ cup borax

Dissolve lye in water, then let cool until lukewarm. Melt grease and cool to lukewarm. Add grease slowly to lye water then

[2]*Ibid.*, p. 146.

add ammonia and borax, and stir until thick. Pour into containers. When it is set (after a few days), cut into squares.

Easy Soap

1 quart cold water	½ cup ammonia
2 quarts grease	2 Tbsp. borax
1 can lye	

Warm grease; add lye, which has been dissolved in water overnight. Stir 15-20 minutes. Add ammonia and borax, which have been dissolved in ½ cup warm water. Stir until thick. Once thickened, pour into molds. Mark the pieces as soon as the soap is cold enough. When it is hard enough (about two weeks), grate it. Store in boxes or plastic bags.

Hand Soap

1 can lye	3 Tbsp. ground oatmeal (optional)
½ cup ammonia	11 cups clear grease
½ cup powdered borax	5 cups soft water
2 ounces lanolin	⅓ cup sugar
4 tsp. aromatic oil of roses (or scent you prefer)	3 ounces glycerine

Measure water into enamel pan. Add to it one at a time until dissolved, stirring constantly: lye (slowly), ammonia, borax and sugar. Continue stirring until cool. Slowly pour in the fat, constantly stirring. Add fragrance and stir 15 minutes. Add lanolin, glycerine, and oatmeal while stirring. By this time the mixture should be thick and creamy. Pour into molds. Allow to stand until solid. Cut into bars. Wrap in wax paper and let stand at least two weeks before using.

Hand Soap No. 2

11 cups grease	½ cup borax
5 cups cold water	½ cup sugar
1 can of lye	4 tsp. oil of sassafras
½ cup ammonia	

Boil sugar in one cup of water and add to rest of water. Mix lye, borax and ammonia, stir while adding the water and until the mixture is cool. Add this mixture to the warm grease, stirring constantly until it begins to thicken. Add the oil of sassafras, stir, and pour into molds. Allow to set up, then cut into bars.

Laundry Soap

5 pounds grease
½ cup ammonia
½ cup powdered borax
1 cup coal oil or kerosene
1 can lye

Dissolve lye in one quart of cold water. Dissolve borax in one cup of water and add lye to mixture. Melt grease and add ammonia and kerosene. Add to lye mixture. Stir until it thickens. Pour into milk cartons or other molds. After it has set up solid and aged at least two weeks, this soap can be ground or grated for laundry soap.

Once you've tried a few of these soaps, you should be able to make your own recipes. Remember, there are only two basic ingredients — lye and fat. Have fun!

30
Be Ready to Sew

This section offers some thrifty ways to apply basic sewing skills. It is not intended to be a course in sewing.

No matter what your situation is at present, it would be beneficial to have at your command enough knowledge of sewing that you could properly clothe your family should circumstances demand it.

During hard times or a crisis, whether it is on a large or small scale, there is some return to home crafts and arts. The ideas presented here should make those problem times (and we all have them) easier and happier.

Supplies and Equipment

When it comes to preparedness, you should have more than just food in the pantry. Your sewing pantry should contain the necessary tools of the trade. A carpenter cannot build a house without the necessary tools; neither can you sew without the proper equipment.

You are the best judge of your needs. One who sews a little doesn't require as many gadgets as one who sews for the whole family and makes slipcovers too.

Notions

It's good to keep on hand a supply of such things as various sizes of elastic, large and small snaps, hooks and eyes, trims, and thread. This will allow you to complete a project easily and properly, whether it is a new one or a repair.

Many new products truly simplify all aspects of sewing. Browse through a notions department in a department store and note what is available. The majority of the items are inexpensive. Build up your supply by adding one or two items a month to your shopping list. You will soon have enough to care for your needs.

Storage of Material

It is wise to have some fabric on hand; not necessarily bolts — but some. Build up your supply by taking advantage of yardage sales, seasonal clearances, and remnant tables. Check remnants for tears, discolorations or imperfections.

Proper care for stored fabrics includes keeping them in dust-proof boxes and labeling them as to the contents. To protect fabrics from insects, mothballs or flaked napthalene can be wrapped in paper and kept in the box. In order to be effective, the box should be airtight.

Mending, Patching, Darning

Do you recall how to mend, patch, or darn? With affluence and advanced technology, for some it has become easier to buy new items than to take the time to repair and extend the life of what they already have. Since affluence might not continue, start now to acquire the skills and knowledge to "make do" with what you have.

Mending Rips and Tears

There are three ordinary types of tears: straight, three-cornered, and diagonal. Mend a tear by stitching a narrow dart on the *wrong* side of the material. The widest part of the dart should be at the outer edge of the cloth, where the tear began. The dart should extend one inch beyond the place where the tear ends. Make the point of the dart very narrow and sharp. When the dart is pressed, the mending will often be scarcely noticeable. Alternatively, mending can be done by machine- or hand-stitching across the tear to reinforce it. The accompanying drawings illustrate ways to stitch the various tears.

Mending tape, one of those marvelous inventions that makes a chore easier, comes in inexpensive kits with various color tapes. A piece of this tape is applied to the *wrong* side of the tear and is permanently pressed on with a warm iron. The tape will stay in place through washing and cleaning. Directions for applying the tape are usually on the package.

Mending Methods

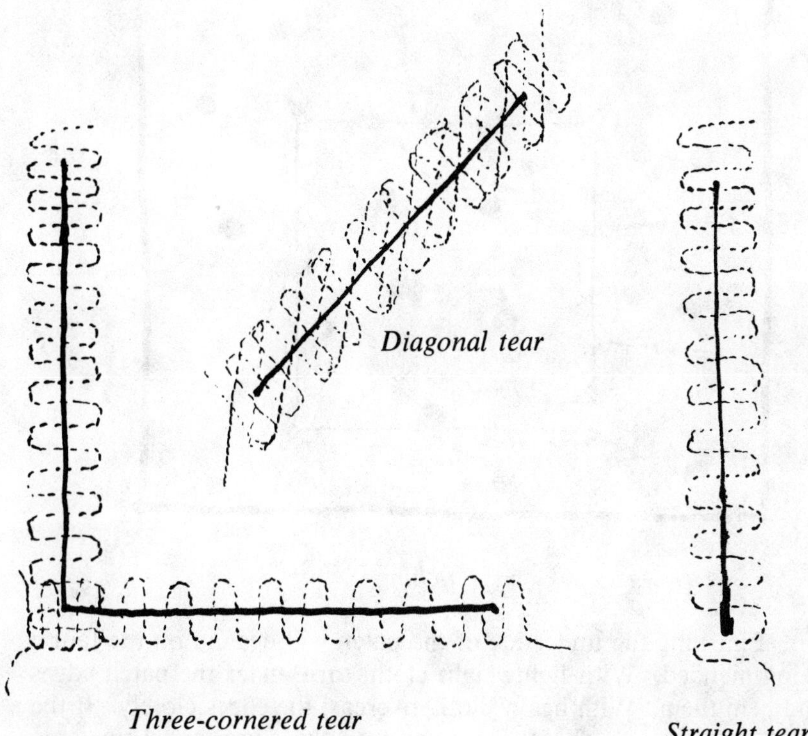

Three-cornered tear

Straight tear

Patches

With care and a little effort, anyone can patch. A hole is generally patched with fabric matching the garment. (Also see "Iron-on Patches" in this chapter.)

— Trim the edges of the hole, making it either oblong or square.

— Clip the corners of the opening one-eighth inch, so that the edges will turn and lie flat.

— Cut a piece of matching material one inch larger on all sides than the hole.

— Place the patch so there is no break in design, then pin patch to underside of the hole.

— Turn under raw edge of the hole, and fasten the patch to the material, using small stitches.

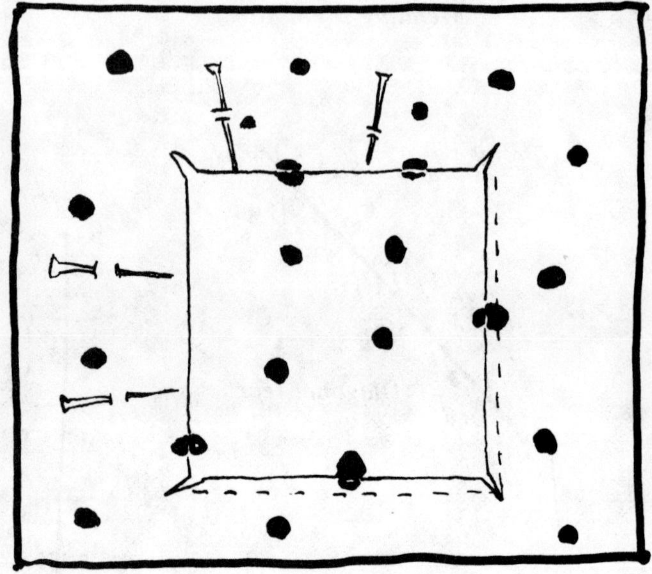

Patching

Finishing the underside of the patch will depend on the fabric being mended. With lightweight cloth, turn under the patch edges and hem them. With heavy cloth, overcast the edges closely. If the patch is large enough, it is easy to turn the edges and hem it on the machine before putting it on. After the patch is sewn on, tack the hemmed edges down.

Iron-on Patches

Iron-on patches, to cover holes in elbows and knees, come in a variety of materials. Denim and corduroy are two that are commonly available.

Decorative appliques and embroidered patches are also available. To use a patch of your own fabric, buy bonding fabric to place between the garment and the patch, then iron it in place. Make interesting patch shapes using cookie cutters or children's coloring book pictures for patterns.

Crochet or embroider flowers or other designs to cover holes in children's wear, adding a few extra motifs for a pleasing effect. I can still remember a favorite maroon, short-sleeved sweater that had flowers and leaves crocheted all over to cover up a few holes

Amusing patches

and prolong its life. The same technique can be used with the embroidered patches sold at the notions counter.

Remaking Clothes

Remade clothes are no longer frowned upon, thanks to a rash of shortages, higher costs of living, inflation, realization that too much is wasted, and budgets in need of balancing. Whatever the reason, it's great! A fine sense of satisfaction comes from increasing the life of a garment, and doing it economically and attractively.

Some of the clothes that have been put in a trunk or the back of the closet because they were too good to throw out can be altered, adjusted, remodeled, remade, or the trim changed; in short, something can be done to give them new life.

To be sure the time and effort will be well spent, examine clothes that might be made over. Is the fabric good enough to warrant the remodeling? Hold the garments up to the light to see if there are holes or thin spots. Perhaps the moths have been busy. Cottons, silks, rayons, and synthetics may have small tears or weak spots.

Clothes only slightly damaged can be salvaged in various ways. Some materials can be turned and used on the wrong side. Perhaps a new pocket, or a bright patch of embroidery can be added to cover up a tiny hole if the rest of the fabric is still good.

It may not be practical to remake the garment in a different style, keeping the same pattern size, but don't overlook the possibility of making children's clothes from larger garments. This can be very successful. One of the most beautiful coats I had as a little girl was made from a man's gray tweed overcoat someone had given to my mother. The rather stark overcoat became a pretty, full coat with a hood lined with wine-red velvet, and it had velvet cuffs. To complete my "new" outfit was a beautiful matching velvet muff.

Taking a Garment Apart

All garments should be washed or dry cleaned before remodeling is started. Rip the seams with a single-edge razorblade — *carefully*. If you have a seam ripper, that's even better. Once the garment is ripped apart, press it and mark the wrong side of the material so that you do not become confused while working on it.

Check Pattern Before Cutting

When you have the material clean, ripped apart and pressed, spread it out flat and place the pattern pieces on it to see if there is enough fabric. Do this *before* you start cutting. If you are short of material, you might have to piece some together or choose another pattern that uses less fabric.

Tips on Remodeling

If the skirt of a dress is too narrow, add a new skirt at the waistline or slightly below in a blending or contrasting color.

Make a vest or jumper from a dress that is worn out under the arms.

A dress can be made larger by inserting a contrasting panel at the front, and perhaps adding matching cuffs, pockets or collar.

Make a playsuit from a summer dress which is too short.

Try making a vest from the tunic top of a pantsuit. Cut it the length desired, open up the front and either insert a zipper or make a facing and use buttons for decoration.

Pantsuit tops or bottoms can always be used as separates. If the top wears out first, make a new top that contrasts or matches, using part of the old one for trim.

Transform mother's skirt into one for daughter.

A bedspread with several worn spots can be converted into a warm bathrobe.

> A synthetic dress can take on new life as a blouse. Here are four kinds of blouses to make (after you cut the dresses, hem the raw edges or finish with a knitted fold-over binding). 1.) Cut the skirt off the dress, leaving a tail, for a blouse to tuck into a skirt. 2.) Make a blouse when you cut off the skirt, leave enough fabric to blouse over to the length you want it plus an allowance for the casing for the elastic. 3.) An A-line dress can make a fitted tunic blouse. Cut the skirt off below the hips, put a couple of rows of casing in the waist to insert elastic. Then wear it outside the skirt, with or without a belt. 4.) Buy ribbing by the yard, add it to the waist (after you cut off the skirt), the cuffs and perhaps the neck to make a blouson or pullover sweater.[1]

Pajama tops and bottoms never seem to wear out at the same time. Don't throw out the jacket if the bottoms are worn; make another pair of pants in a contrasting color.

If you have a full slip with a worn top, cut off the top, make a casing about the waist, insert elastic, and you have a half slip.

Remodel a man's shirt into a blouse for a woman or little girl. It might have a frayed collar or cuffs and buttons missing, but you can work magic with it. Dad's shirt will yield enough fabric for a little girl's dress or slip. Also from a shirt could come a boy's shirt or a child's sunsuit or overalls.

It is often possible to cut a girl's dress from the skirt of one of mother's or big sister's dresses.

---

[1]"How to Look in Your Closet and Find a New Wardrobe," *Woman's Day* (Greenwich, Connecticut: Fawcett Publications, 1974), p. 51.

Try using white or a contrasting color for a new yoke, sleeves, collar, pockets or cuffs.

When you remodel a coat by changing the collar and cuffs, a piece of fake fur does wonders.

For remodeling a man's suit, try making a boy's suit, a child's coat, a short jacket. If it is used for a girl's jacket, trim the collar, cuffs and pockets with velvet or fake fur.

Long pants can be cut off for shorts when the knees wear through; that is, if there is no one younger for whom the pants should be patched.

Salvage Usable Items

Here is one more way to stretch that budget. After you have sorted through your wardrobe and decided what to remodel, you might find one or two items that will have to go. Don't throw them out yet. They have parts that are still in excellent condition.

Buttons. Remove all the buttons. By taking two minutes to string them together, it will be easy to find the set. A button box is good to have for repair jobs and for making new items.

Zippers. No need to throw away a perfectly good zipper. Carefully remove it with a razor blade or seam ripper, pull out the broken threads, then roll it up and tuck it away.

Trims. Save the trims or special appliques that are still in good condition.

Remove brocade and embroidered trims, lace, and fake furs. At the very least, you can use them to trim doll clothes.

Snaps, hooks and eyes. Since all of these items cost money, and old ones function the same as new, why not use them?

The garment is now stripped of everything useful, but don't throw it out yet. Put any good pieces of fabric in the scrap box, then use what's left for dust cloths.

Scrap Box

A scrap box is a place to keep any good-sized fabric remnants. It can save you money because it takes only a small piece of material to make collars, cuffs or pockets. Use larger scraps for shirt or dress yokes, and make a pocket to match.

Make patches, either conventional or novelty style, from the scrap box.

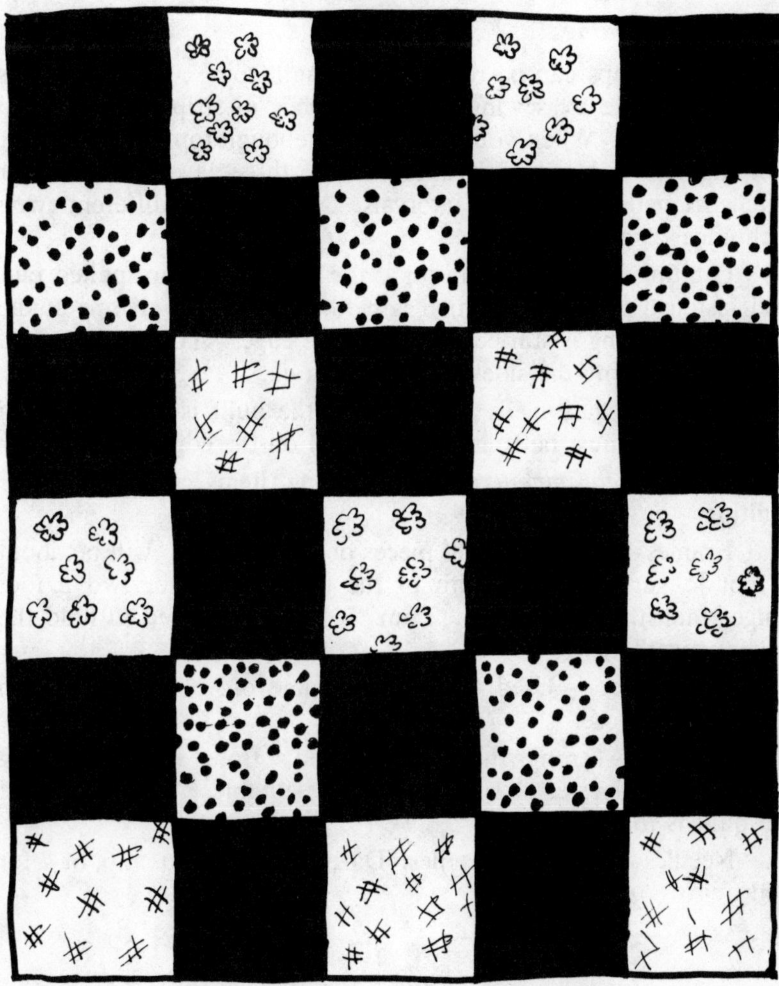

Quilting

Larger scraps would be enough for tiny baby jackets.

Aprons made from remnants or leftovers can be finished with contrasting bands, pockets or ruffles.

Pot holders can be easily made from scraps.

A patchwork-style shoe bag, using different fabrics for each pouch or section, would be clever.

A good scrap box can be very useful. Don't let your scrap box become a catchall or a rag box.

Quilts and Bedding

Some scraps can be recycled into quilt tops. Patchwork tops are easily made by sewing pieces together in strips or blocks to form a pattern. When you have a large enough top, it is laid over a lining and a layer of batting. These three layers are tied or stitched together at spaced intervals. A number of different tying techniques can be used.

Quilting is easier and faster if the three layers are pulled taut and fastened to a frame which is held rigid with C clamps at the corners. Quilting is started at the outside edges. You work toward the center, rolling the sides under as you go.

After the quilting is finished and the quilt is removed from the frames, it must be finished or bound on the edges.

Quilting equipment. The following items are needed for quilting.

Frames — Four 1″ x 3″ pieces of lumber. Two will be about a foot longer than the width of the quilt; the other two a foot longer than the quilt length. Four C clamps are used to hold the frames together.

Saw horses — Used to rest the frame on. Chairs may also serve this purpose.

Thread — Strong thread is needed. Both special quilting thread and polyester sewing thread are suitable. Yarn is used if the quilt is to be tied.

Needles — Quilting needles. Darning needles no. 5, 6, or 7 for tied quilts.

31

Coping with Emergencies

None of us look forward to problems with glee. In doing ordinary things, most of us come close to disaster at one time or another. We are usually on the safe side, but when a broken ankle comes, a kitchen fire, a shattered picture window, or a falling ladder, suddenly it's a different world. We need help.

Can you manage until help comes? Can you manage without help? What are your urgent needs? At least thinking about the pros and cons of a situation before it arrives will help you to cope with it.

One solution is to prevent accidents whenever possible. Use common sense and follow safety rules in the home, outside, and at work.

Even when we take precautions, problems still come at times. While no one knows what emergency situation might come, the best advice is to learn to act and not react.

> You must survive until safe. So after sustaining life through first aid, *stop.* Try to figure out what kind of a jam you are in. Relax, if only for a few minutes, and try to fight off any feelings of panic. Saying a prayer might not be a bad idea. Keep first things first.[1]

With realistic action, you can prevent tragedy from the dangers that lurk around every corner. Get the prevention rules firmly in your mind. Obtain a Red Cross or similar first-aid book and keep it in an easily accessible place.

---

[1]*Fieldbook for Boys and Men* (New Brunswick, New Jersey: Boy Scouts of America, 1967), p. 303.

It is wise to have on hand some medical and first-aid supplies to help cope with the small-scale emergencies. The list of such supplies must be tailored to suit the needs of each individual family. The basic list below is recommended as a start for everyone.

LIST OF BASIC SUPPLIES

Antiseptic solution
Aspirin tablets (5 grain)
Baking soda
Cough mixture
Diarrhea medication
Disinfectant
Ear drops
Table salt
Toothache remedy
First-aid handbook
Pregnancy supply
Specific medications (recommended by your doctor)
Adhesive tape, roll (2″ wide)
Applicators, sterile, cotton tipped
Bandage, sterile roll (2″ wide)
Bandage, sterile roll (4″ wide)
Bandages, triangular (37″ x 37″ x 52″)
Band-aids (assorted sizes)
Cotton, sterile, absorbent
Laxative
Motion sickness tablets
Nose drops
Petroleum jelly
Rubbing alcohol
Smelling salts
Dressings, sterile (4″ x 4″)
Hot water and enema bag
Medicine dropper
Safety pins, assorted sizes
Sanitary napkins
Soap
Scissors
Splints, wooden (18″ long)
Thermometer
Tweezers
Water purification materials

Care and Maintenance of Supplies

Medicines obtained for your emergency supply should be labeled so that the name of the medicine, instructions for use, and necessary warnings, such as "For external use only" and "POISON," are clearly visible. These medicines should be carefully packed to prevent breakage, and stored in a dry, cool space out of reach of children. The best storage temperature is below 70° F., but they should not be frozen. Some medicines may have to be replaced periodically. If it is a medicine that is used regularly, be sure it is used before the expiration date.

Emergency Toilet Facilities

A person who went through a Midwest flood told me that one of the greatest problems was sanitation. Accustomed to the convenience of modern facilities, people had made no emergency provisions, and this had a devastating effect on morale. Conditions rapidly became deplorable.

In like manner, if you do not prepare ahead of time, a minor emergency could turn into a nightmare. It is vital that some way be devised to dispose of human wastes. Various conditions might arise that would prevent use of the bathroom. For example, water may not be available to use in flush tanks or basins. Failure to *properly* remedy the situation can lead to the rapid spread of germs and disease.

It is advisable to keep on hand some sanitation supplies, such as:

— A metal container with a tight fitting lid, to use as an emergency toilet. This could be fitted with some kind of seat. An old toilet seat kept specifically for this purpose would be excellent.

— A larger container, also with a tight-fitting cover, should be used to empty the contents into for later disposal.

— Plastic bags to be used as can liners. They would facilitate disposal of wastes and help to keep odors at a minimum.

— A supply of old newspapers and grocery sacks would be useful for wrapping garbage or lining waste containers.

— A reserve of toilet tissue, soap, and feminine hygiene items should be stored.

— A disinfectant such as chlorine bleach.

If it is possible, bury the waste and garbage in a hole one to two feet deep. This depth is necessary to prevent dogs from digging it up, and to reduce the possibility of insects or rodents spreading germs and disease.

Each time the temporary toilet is used, pour or sprinkle a disinfectant such as chlorine bleach or quicklime into it. This will help keep down germs and odors.

Individual privacy is important. Screen temporary toilet facilities from view by hanging a blanket, sheet, canvas, or tarp.

Be aware of the needs of an infant in the home, and store items such as diapers, extra blankets, and plastic pants.

32
Storage Space

When you consider buying extra supplies of anything, the first and most logical question is: Where will I put it? This is the question that discourages many people from starting an emergency preparedness program. But don't let this problem discourage you. There are many ways to solve it.

Small Home or Apartment

— Build shelves and hang a curtain in front of them.

— Put items in boxes and stack them under beds, cribs and tables.

— Decorate or cover large drums or trash cans.

— Instead of using bricks to build a bookshelf, use no. 10 cans filled with supplies.

— Place the sofa twelve to fifteen inches from the wall, then stack boxes behind it.

— Keep an accurate up-to-date list of where you have put things.

Under the Kitchen Sink

If your kitchen sink area contains a lot of dead space, build shelves for storing cleaning and laundry supplies.

Partial side shelves may be held in place with brackets. Two full-length shelves supported with cleats, or any of several other items can be fastened to the inside of the doors for convenient storage. Possibilities to consider include: towel rods, paper towel holder, wastebasket, garbage can, dish drainer, shelf for cleaning supplies and a sink rack.[1]

Pantry

Built-in pantries are fine, but does yours waste space? Is there too much room between shelves? Organize your cans and boxes so that one shelf holds the same size item. If you stack cans or boxes more than two high, you'll spend too much time restacking fallen goods. If the shelves are adjustable, arrange them so there is a space of two inches between the top of the cans and the bottom of the next shelf.

Don't make the shelves too deep. Getting to items in the back can become so difficult that they may stay there until they are spoiled. Deep shelves make it difficult to rotate your supplies. Put the heaviest items at the bottom, those items most often used at a height between the hips and shoulders, and paper items on the top shelves, stacked right to the ceiling if necessary. If they should fall, no one will get hurt.

Shelves

In many bedroom closets there is only one shelf, but room for at least two — sometimes more. Use this space. A piece of one-by-twelve-inch pine makes good shelf material.

If you are renting an apartment and feel you can't attach anything to the floor or walls, use boxes stacked on their sides. When I was first married, we lived in a small rented home. In the bedroom we later used as a nursery, we stacked apple boxes, one on top of the other, for storage shelves. On the back sides of the boxes nearest the bedroom area we tacked up peg board, painted it to match the nursery furniture and hung pictures and toys on it for decoration. We placed the children's dresser in the closet — none of their clothes hung very low so the space at the bottom of the closet otherwise would have been wasted. A curtain hung from the wall to the boxes closed off the doorway to this small storage room. There wasn't much room in the nursery, but enough for a storage room.

---

[1]Rhea H. Gardner, *More Storage Space for Your Kitchen* (Logan, Utah: Utah State University Extension Service), p. 6.

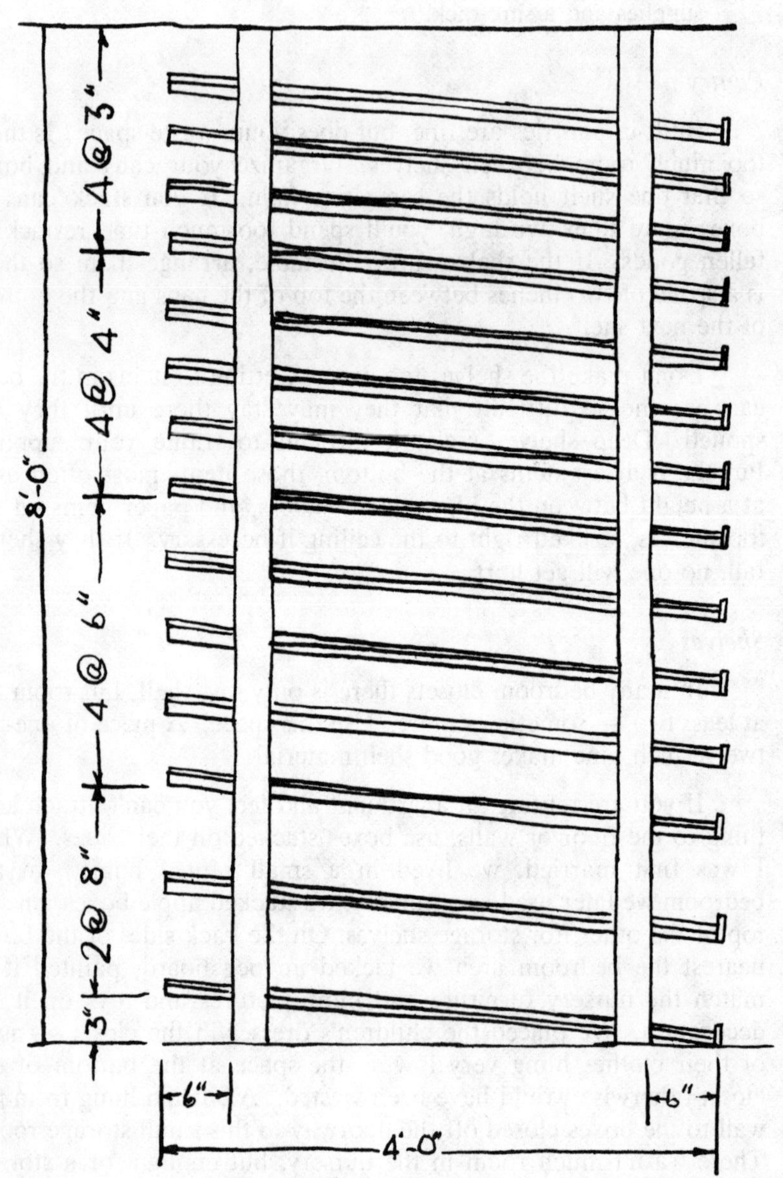

Slanting shelves

Another time, as a place to store canned goods (which need to be rotated at least every six months), my husband built shelves with one end slightly higher than the other. I put newly purchased cans on the high end of the shelf (on their side) and took older ones from the low end. Each time I removed a can, the whole stack rolled over one-half turn. No more turning individual cans by hand!

This type of shelf can be free standing or it can be built against a wall. If it is free standing, make it as wide as desired. If it is built against a wall, it should be no more than sixteen inches wide. Since it is loaded from the front, a shelf wider than this becomes difficult to load.

To prepare to make a shelf of this type, measure the height of the cans to be stored, to determine how many cans laid end to end will fit on the shelf width, leaving ¾ inch between each can. In this space between rows of cans, tack down a piece of ¼-x-½-inch molding to separate the cans. Glue and tack a piece of ½-x-¾-inch wood across the lower edge of each shelf to keep the cans from rolling off onto the floor. For storing up to eight hundred cans of assorted sizes, use ⅜-inch plywood 4 x 8 feet for shelves and cut six shelves from each piece.

Attic

> Poor ventilation and a wide variation in temperature limits the value of the attic as a storage place. Crossventilation will help to prevent extreme heat, but it is difficult to keep the attic warm in very cold weather and cool in hot weather. . . . Popcorn, herbs, and dried seeds can remain there indefinitely.[2]

Basement

> A basement room that is dry all seasons of the year, free from steam, hot air and water pipes, and heat of any kind, and one which is closed off from the rest of the house is an ideal storage room. Under these conditions the average yearly temperature will be between 50 to 60 degrees F., the temperature at which most foods will retain their maximum keeping qualities.[3]

---

[2]A. L. Weaver, *Winter Vegetable Storage,* Circular 530 (Urbana, Illinois: College of Agriculture).

[3]Celectia J. Taylor, *Relief Society Magazine* (Salt Lake City, Utah, August 1956), p. 462.

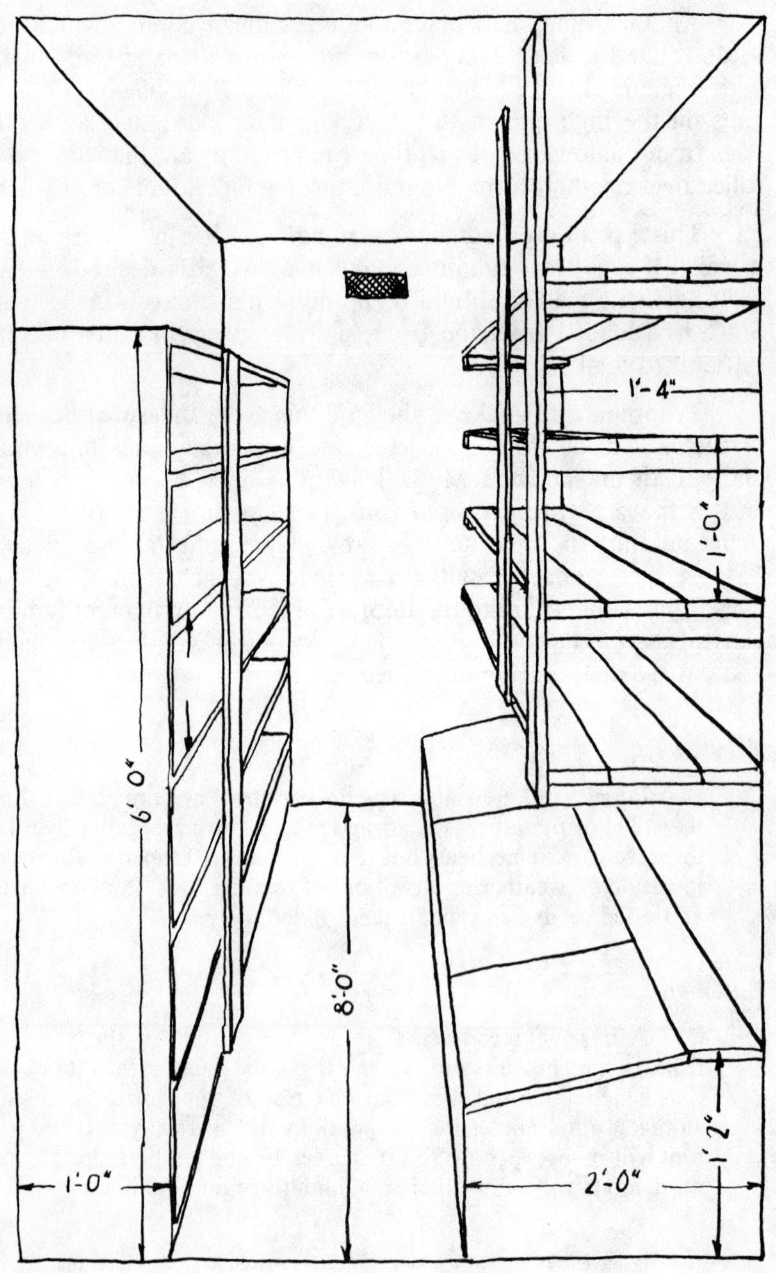

Storage room

The size of the storage will depend upon the food habits and likes of the family. A storage room ten feet square and seven feet high, properly arranged, is large enough for a family of eight to ten people. Where space is limited, a smaller room can be arranged to provide adequate storage for a large family.[4]

Garage

A garage could provide ample storage space. My husband and I once purchased a home without a basement, and the garage became our storage area. An area eight feet by twelve feet was measured off and enclosed with walls and ceiling. After installing shelves from floor to ceiling, there was more than enough room for storage and there was still room for the car.

Insulation and Ventilation

Wherever you place your supplies, make sure that the room is properly insulated, and has good crossventilation. Supplies can be stored much longer under these conditions. The storage room inside our garage was insulated in both walls and ceiling. Even the door was insulated. And a fan was installed to insure adequate air circulation.

---

[4]Bulletin N.S. 148, *Home Storage in Utah* (Logan, Utah: Utah State University, 1969).

33
How Much Is Enough?

This chapter is designed to help you apply to your own circumstances the ideas presented in the preceding chapters. Use the detail in this chapter to draw up worksheets. Remember that, to be of any value, the decisions you make must be related to *your* family needs.

To help you determine long-term food needs, you may want to first measure short-term consumption. Make a chart headed "Food Eaten from . . . . to . . . ." In the left-hand column list line by line the principal categories of food the family eats: Milk, cheese, butter, eggs, breads, cereals, vegetables, fruits, meats, flour, sugar, condiments, and so on. Have a column for each day of the week. Tape the chart to your refrigerator door, and note on it each time a food item is used up. Use your own method of marking. Example:

Items	Sun.	Mon.	Tues.	Wed.	Thurs.	Fri.	Sat.
Milk (half-gal.)	//	/	//				
Eggs	2	4 + 2	4				

Within two or three weeks you can get a reading that will assist you in longer-term planning.

In that planning, as in daily usage, keep firmly in mind the Basic Four food groups (see chapter 1). Keep in mind too the need to know what you already have on your family "inventory." With pencil in hand, make an accurate survey of the foods and sundries you have on hand. You could use a format such as this:

INVENTORY — WHAT DO I HAVE?

Fruits and Vegetables		Cereals and Breads		Proteins — Meat		Dairy Products — Milk	
Item	Amount	Item	Amount	Item	Amount	Item	Amount

Next, write down the items you consider necessary, in order of their priority, using a planning sheet such as this:

PLANNING SHEET FOR . . . (Period of Time)

Item	Amount Needed for 1 x No. in Family =	Amount on Hand	Amount to be Obtained
	x =		
	x =		

This process will take time, but don't be discouraged. It is time well spent.

You will perhaps need a reminder list of staple food items. Use the following larder list as a starter, then add to it as necessary.

LARDER LIST

Beverages
Cereals, grains, breakfast foods, spaghetti, macaroni, noodles
Fats and oils for cooking purposes
Milk
Eggs
Flour — a variety
Sugar — brown, confectioners, granulated
Butter, margarine
Puddings, gelatins
Cheeses
Breads
Juices — vegetable, fruit
Fruit — a good variety: canned, dried, fresh, frozen
Vegetables — a good variety: canned, dried, fresh, frozen
Salad dressings
Condiments, relishes, catsup, mustard, pickles
Canned meats
Canned fish
Soups
Sauces — Tomato
Baking powder, baking soda, cornstarch
Salt, pepper, spices and herbs
Flavorings
Bouillon
Peanut butter

Similarly, here is a partial list of nonfood items.

NONFOOD ITEMS

Adhesive tape
Alcohol, rubbing
Ammonia
Aspirin
Baby needs
Bandages
Batteries
Blankets, bedding
Boric acid
Candles
Cleaning supplies
Clothespins and rope
Clothing
Cold remedies
Cooking needs
Dish soap
Disinfectants
Epsom salts
Egg beater, manual
Feminine hygiene items
First-aid book, kits (kits can be made at home, kept in one place)
Flashlight and fresh batteries
Foil, waxpaper, etc.
Gauze
Grinder
Hot water bottle
Iodine
Kaopectate
Kettles
Knives
Light bulbs
Matches
Milk of Magnesia
Mouse traps
Paper items
Pencils
Rags, clean
Razor blades
Rope
Salt
Seeds
Sewing supplies
Shampoo
Soap—laundry, hand
Soda
Tea (for tannic acid)
Toilet tissue
Tools
Toothbrushes
Toothpaste
Vaseline
Washboard

Use a "think sheet" to organize your home preparedness needs into categories which you can then break down into specifics. As a starter, here are some suggested headings of categories:

Foods and food preparation
Sewing needs
Books
Cleaners, soaps, washing facilities
Fuel and light
Living and camping equipment
Medicine/Hygiene — personal
Seeds
Linens

With the types of needs established, your refrigerator-door chart contributing its reading, and your inventory taken, your long-term planning is taking shape. In computing needed quantities it may help to consult the accompanying table, "Number of Servings from a Unit Measure." You may also find useful the table, "Suggested Quantities of Food in Monthly and Yearly Amounts." Take into account too the expected shelf life of any food you may have on hand. For some of the common foods, see the list of shelf lives.

NUMBER OF SERVINGS FROM A UNIT MEASURE

Food	Unit Measure	Number of Servings
Bread	1 lb. loaf	19 ave. slices
Butter or margarine	1 lb. (2 cups)	48 pats
Cheese		
American or brick	1 lb. (2⅔ cups, cubed 4 cups, grated)	14-16 slices
Cottage	1 lb.	6
Cocoa	1 lb. (4 cups)	100
Crackers	1 lb. (small, square)	108 crackers
Farina (other granular cereals)	1 lb. (3 cups)	25
Fruits		
Apples	1 lb.	3-4
Apricots		
Fresh	1 lb.	8-12
Dried	1 lb.	12
Bananas	1 lb.	3
Berries	1 qt.	4-6
Cherries	1 qt. (2 cups pitted)	6
Peaches		
Fresh	1 lb.	4
Dried	1 lb. (3 cups)	12
Pears	1 lb.	4
Pineapple	2 lb.	6
Prunes (dried)	1 lb. (2½ cups)	8

Food	Unit Measure	Number of Servings
Gelatin	3 oz. package	4-6
Macaroni, spaghetti, or noodles	1 lb.	16-20
Meat		
Boned or ground	1 lb.	3-4
Large amount of bone	1 lb.	1-2
Medium amount of bone	1 lb.	2-3
Oats, rolled	(5 cups)	12
Rice	1 lb. (2¼ cups)	12
Vegetables (dried)		
Beans		
Kidney	1 lb. (2⅓ cups)	12
Lima	1 lb. (2⅓-3 cups)	10
Navy	1 lb. (2-2½ cups)	10
Peas (split)	1 lb. (2 cups)	10
Vegetables (Fresh)		
Carrots	1 lb.	5
Onions	1 lb. (3 large)	4
White potatoes	1 lb. (3 medium)	3
Sweet potatoes	1 lb. (3 medium)	3

SUGGESTED QUANTITIES OF FOOD IN MONTHLY AND YEARLY AMOUNTS

Family Members	Milk		Eggs		Meats		Citrus Fruits Tomatoes		Leafy Green, Yellow Vegetables		Potatoes		Beans Peas Nuts		Other Fruits Vegetables		Flour Cereal		Fats Oils	
	Mo. Qts.	Yr.	Mo. No.	Yr.	Mo. Lb.	Yr. Oz.	Mo. Lb.	Yr. Oz.	Mo. Lb.	Yr. Oz.	Mo. Lb.	Yr. Oz.	Mo. Lb.	Yr. Oz.	Mo. Lb.	Yr. Oz.	Mo. Lb.	Yr. Oz.	Mo. Lb.	Yr.
Children 1-6 years	24.0	312.0	20.0	260.0	3.2	41.6	8.96	58.2	8.96	58.2	4.0	52.0	0.8	10.4	5.6	72.8	7.2	93.6	2.0	26.0
7-12 years	24.0	312.0	24.0	312.0	7.2	93.6	9.6	124.8	8.0	104.0	11.2	145.6	2.0	26.0	4.48	58.2	9.6	124.8	½ lb. butter or margarine per person per week	
Girls 13-20 years	24.0	312.0	24.0	312.0	8.0	104.0	9.6	124.8	12.0	156.0	12.0	156.0	2.0	26.0	4.48	58.2	15.2	197.6		
Boys 13-20 years	28.0	264.0	24.0	312.0	8.0	104.0	11.2	145.6	8.48	110.2	19.2	249.6	3.2	41.6	11.2	145.6	20.0	260.0	4.0	52.0
Women	20.0	260.0	24.0	312.0	8.0	104.0	9.6	124.8	9.6	124.8	12.0	156.0	2.0	26.0	4.48	58.2	15.2	197.6	1 lb. cooking fat or oil per week for family of 4-5	
Men	20.0	260.0	24.0	312.0	8.0	104.0	9.6	124.8	11.2	145.6	16.0	208.0	2.4	31.2	11.2	145.6	20.0	260.0		
Amount needed for your family																				

Remember these amounts are only suggested. Any area can be increased or decreased as long as the overall diet remains balanced. Amounts will vary according to usage in various families. Use these figures to estimate the food needs of your family for one month to a year.

SHELF LIFE OF COMMON FOODS

(Note: Store foods in coldest cupboards — away from heat. While foods will be safe beyond the recommended storage times, flavors and textures deteriorate. Date foods, then use up oldest foods first.)

Food	Length of Storage at 70° F.
Staples	
Baking powder	18 months — or expiration date on the can
Baking soda	2 years
Bouillon cubes or granules	2 years
Cereals — ready-to-eat	6-12 months
Cocoa mixes	8 months
Cornmeal	12 months
Cornstarch	18 months
Gelatin	18 months
Salad dressings (unopened)	10-12 months
Salad oil (unopened)	6 months
Vinegar	2 years

Food	Length of Storage at 70° F.
Mixes and Packaged Foods	
Biscuit, brownie, muffin	9 months
Cake mixes	1-2 months
Pancake mix	6-9 months
Pudding mix	12 months
Sauce and gravy mix	6-12 months
Soup mix	12 months
Condiments	
Catsup, chili sauce (unopened)	12 months
Mustard, prepared (unopened)	2 years
Vanilla	2 years
Other extracts	12 months
Other	
Coconut (unopened)	12 months
Instant Breakfast	6 months
Whipped topping (dry)	12 months

Information based on *Spotlight on Cupboard Storage* by USDA and Ohio State University, 1973.

As well as *what* to buy, for many foods *when* to buy is significant. See the table below.

MOST ECONOMICAL TIMES TO BUY

1. Meats
   - Beef — Early in year
   - Veal — Last half of year
   - Lamb — Winter
   - Pork — Autumn
   - Turkey — Last half of year
   - Chicken — Broilers: November, December, January
Stewing: Fall and winter
Fryers: Spring
2. Vegetables — During the growing season / Home grown or from producer
3. Fruits — During the peak of season / Home grown or from producer
4. Honey — August and September / From producer
5. Wheat — In the fall / Direct from the farmer or mill
6. Beans — September and October / From the grower
7. Sugar — Discount sales are usually during the winter

PLANNING GUIDE FOR HOME-PRESERVED FOODS

Product	Number of Times Served	Approximate Size Serving	Amount Needed One Person	Amount Needed Family of Four
Citrus fruit and tomatoes	7 per week 52 weeks	1 cup	91 quarts	364 quarts
Dark green and yellow vegetables; broccoli, spinach, greens, carrots, pumpkin, sweet potatoes, yellow squash	4 per week 52 weeks	½ cup	26 quarts	104 quarts
Other fruits and vegetables; apples, apricots, peaches, pears, asparagus, green beans, corn, peas	17 per week 52 weeks	½ cup	109 quarts	436 quarts
Meat, fish, poultry	4 per week 52 weeks	½ cup	26 quarts 52 pints	104 quarts 208 pints
Jams, jellies, preserves	6 per week 52 weeks	2 Tbsp.	40 half pints	160 half pints
Relishes	3 per week 52 weeks	1 Tbsp.	5 pints	20 pints
Pickles	4 per week 52 weeks	——	26 pints	104 pints

This guide is based on the USDA *Daily Food Guide*. By using this guide, adjusting for likes and dislikes, need and appetites, you should be able to approximate your goal for a season's canning, freezing, or drying.

Finally, if you can, preserve foods at home. The accompanying planning guide will help you to calculate your needs in this respect.

How much is enough? As regards *your* needs, only you can tell. If this chapter helps you in that decision, and if from this book you gain the know-how and the motivation to supply the basic needs of yourself and your family for now and for the future, the book will have achieved its aim.

Bibliography . . . .

ARTICLES

Egan, Merritt H., M.D. "Home Storage Advice Prepared and Compiled," February 1959.

Graham, Jewel. "Food and Nutrition, What Makes a Good Buy?" Ames, Iowa: Iowa State University Extension Service, 1967.

Gunther, Max. "The Era of Substitute Foods," *Good Food,* Radnor, Pennsylvania: Triangle Publications, January 1974.

Hellman, Hal. "The Story Behind Those Meatless Meats," *Popular Science,* October 1972.

Howe, Allie, ed. "Building a Food Storage Room," *Improvement Era,* Salt Lake City, The Church of Jesus Christ of Latter-day Saints, July 1956.

———. "Planning and Caring for Food Storage," *Improvement Era,* Salt Lake City, The Church of Jesus Christ of Latter-day Saints, August 1956.

Nutrition the Name of the Game. Marshall, Minnesota: The Egg Products Division of Marshall Foods, Inc.

Naylor, Esther C. "Storing Food in a Two-Room Apartment," *Relief Society Magazine,* Salt Lake City, The Church of Jesus Christ of Latter-day Saints, August 1948.

Nicholes, Dr. and Mrs. Henry J. *Home Drying of Fruit and Vegetables.* Compilation.

Simonsen, Velma N. "Storing Food in a Basement," *Relief Society Magazine,* Salt Lake City, The Church of Jesus Christ of Latter-day Saints, 1948.

"Take a Fresh Look at Eggs," *Good Foods,* Radnor, Pennsylvania: Triangle Publications, 1974.

The National Dairy Council. "Food Faddism," *Dairy Council Digest,* Vol. 44, No. 1, January-February 1973.

University of California at Davis Extension Service. *Home Storage of Nuts, Cereals, Dried Fruits and Other Dried Products,* Davis, California, 1969.

Pamphlets

Bardwell, Flora. *How to Cook and Use Whole Kernel Wheat,* Logan, Utah: Utah State University Extension Service, 1973.

———. *Rice,* Logan, Utah: Utah State University Extension Service, 1968.

Bulgur Wheat Recipes. Logan, Utah: Utah State University Extension Service, 1973.

California Vegetable Concentrates Product Calculator, Modesto, California: General Foods Corp., 1973.

Clemm, Mary Lou. *Wix and Wax, the Complete Book of Candle Making,* Edgewood, Maryland, 1974.

Creating Candles, Temple City, California: Craft House Publishers, Inc., 1970.

Food Protein Council. *The Protein Power of Soybeans,* Minneapolis, Minnesota: General Mills, 1973.

"Honey — The Collectors Cookbook," *Woman's Day,* Greenwich, Connecticut: Fawcett Publications, July 1968.

Miller, Elna. *Facts About Food and Nutrition,* Logan, Utah: Utah State University Extension Service, 1973.

———, and Knowlton, George. *Food Storage in the Home,* Logan, Utah: Utah State University Extension Service, 1973.

National Soybean Processors Association. *The Story of Soy Protein,* Washington, D.C.: Food Protein Council, 1973.

North Dakota State University Cooperative Extension Service. *Beans, Beans, Beans,* Circular HE119, Fargo, North Dakota, 1973.

———. *The Goodness of Wheat,* Circular HE123, Fargo, North Dakota, 1973.

Ohio State University Extension Service. *How to Dry Fruit at Home,* USDA and Ohio State University, 1973.

———. *Spotlight on Cupboard Storage,* USDA and Ohio State University, 1973.

———. *Spotlight on Freezer Storage,* USDA and Ohio State University, 1973.

———. *Spotlight on Refrigerator Storage,* USDA and Ohio State University, 1973.

The American Spice Trade Association. *A Guide to Spices,* Technical Bulletin 190, Englewood Cliffs, New Jersey.

USDA. *Cereals and Pastas in Family Meals,* Home and Garden Bulletin No. 150, Washington, D.C.: U. S. Government Printing Office, 1968.

———. *Eggs in Family Meals,* Home and Garden Bulletin No. 103, Washington, D.C.: U. S. Government Printing Office, 1973.

———. *Family Food Budgeting,* Home and Garden Bulletin No. 94. Washington, D.C.: U. S. Government Printing Office, 1971.

———. *Family Food Stockpile for Survival,* Home and Garden Bulletin No. 77, Washington, D.C.: U. S. Government Printing Office, 1966.

———. *Freezing Meat and Fish in the Home,* Washington, D.C.: U. S. Government Printing Office, 1973.

———. *Home Freezing of Fruits and Vegetables,* Home and Garden Bulletin No. 10, Washington, D.C.: U. S. Government Printing Office, 1951.

———. *Honey, Some Ways to Use It,* Home and Garden Bulletin No. 37, Washington, D.C.: U. S. Government Printing Office, 1953.

———. *How to Buy Beans, Peas, and Lentils,* Washington, D.C.: U. S. Government Printing Office, 1970.

———. *How to Buy Eggs,* Washington, D.C.: U. S. Government Printing Office, 1970.

———. *How to Buy Fresh Vegetables,* Washington, D.C.: U. S. Government Printing Office, 1970.

———. *Keeping Food Safe to Eat,* Home and Garden Bulletin No. 162, Washington, D.C.: U. S. Government Printing Office, 1973.

———. *Nutrition, Food at Work for You,* Washington, D.C.: U. S. Government Printing Office, 1971.

———. *Nutritive Value of Foods,* Home and Garden Bulletin No. 72, Washington, D.C., U. S. Government Printing Office, 1971.

———. *Protecting Woolens Against Clothes Moths and Carpet Beetles,* Washington, D.C.: U. S. Government Printing Office, 1971.

———. *Simplified Clothing Construction,* Home and Garden Bulletin No. 59, Washington, D.C.: U. S. Government Printing Office, 1967.

———. *Storing Vegetables and Fruits in Basements, Cellars, Outbuildings and Pits,* Home and Garden Bulletin No. 119, Washington, D.C.: U. S. Government Printing Office, 1966.

Utah State University Extension Service, *Home Drying of Fruits and Vegetables,* Logan, Utah: USDA and Utah State University, 1973.

———. *More Storage Space for Your Kitchen,* Logan, Utah, 1968.

———. *Vegetable Garden Guide,* Logan, Utah, 1974.

The Vacu Dry Co. *Low Moisture Fruits, Their Uses and Advantages,* Emeryville, California, 1962.

Vegetable Protein Foods, Worthington, Ohio: Worthington Foods, Inc., 1973.

Books

Abraham, George. *The Green Thumb Book of Vegetable Gardening,* New York: Dell Publishing Co., 1973.

Arlin, Marian T. *The Science of Nutrition,* New York: Macmillan Co., 1972.

Ball Blue Book, El Monte, California: Ball Brothers Co., Inc., 1973.

Batchelor, Walter. *Gateway to Survival,* Layton, Utah, 1972.

Boy Scouts of America. *Fieldbook,* New Brunswick, New Jersey, 1967.

Charley, Helen. *Food Science,* New York: The Ronald Press Company, 1970.

Cruess, William V. *Home and Farm Food Preservation,* New York: Macmillan Co., 1918.

Department of Defense. *In Time of Emergency,* Washington, D.C.: U. S. Government Printing Office, 1971.

———. *Personal and Family Survival,* Washington, D.C.: U. S. Government Printing Office, 1966.

Dickey, Esther. *Passport to Survival,* Salt Lake City: Bookcraft, 1969.

Dickson, Sally, and Blondin, Frances. *The New Encyclopedia of Modern Sewing,* New York: The National Needlecraft Bureau, 1943.

Doty, Walter, ed. *All About Vegetables,* San Francisco: Chevron Chemical Co., 1973.

Flack, Dora. *Fun with Fruit Preservation,* Bountiful, Utah: Horizon Publishers, 1973.

Girl Scouts of America. *Cooking Out of Doors,* New York, 1946.

———. *Girl Scout Handbook,* New York, 1953.

Hershoff, Evelyn Glantz. *It's Fun to Make Things from Scrap Materials,* New York: Dover Publications, Inc., 1964.

Hughes, Osee, and Bennion, Marion. *Introductory Foods,* New York: Macmillan Co., 1962.

Johnson, Theta. *Basic Clothing Construction,* Logan, Utah: Utah State University Extension Service, 1974.

Justin, Margaret M., Rust, Lucile R., and Vail, Gladys E. *Foods,* Massachusetts: Houghton Mifflin Co., 1948.

Kains, M. G. *Food Gardens for Defense,* New York: Grossett & Dunlap, 1942.

Kerr Home Canning Book and How to Freeze Foods, Sands, Oklahoma: Kerr Glass Manufacturing Corp., 1974.

Lynch, Mary. *Sewing Made Easy,* Garden City, New York: Doubleday, 1952.

Morse, Sidney, *Household Discoveries, An Encyclopedia of Practical Recipes and Processes,* New York: The Success Co., 1908.

Potter, M.D. *Fiber to Fabrics,* New York: McGraw-Hill, 1953.

Rathbone, Lucy, and Tarpley, Elizabeth. *Fabrics and Dress,* Cambridge, Massachusetts: The Riverside Press, 1943.

Salsbury, Barbara G. *Just Add Water,* Bountiful, Utah: Horizon Publishers, 1972.

———. *Tasty Imitations,* Bountiful, Utah: Horizon Publishers, 1973.

Sherman, Henry C., and Lanford, Caroline Sherman. *Essentials of Nutrition,* New York: Macmillan Co., 1947.

Silver, Fern. *Foods and Nutrition,* New York: D. Appleton-Century Co., Inc., 1941.

Thomas, Dian. *Roughing It Easy,* Provo, Utah: BYU Press, 1974.

USDA. *Composition of Foods: Raw, Processed, Prepared,* Agriculture Handbook No. 8, Washington D.C.: U. S. Government Printing Office, 1963.

———. *Food, The Yearbook of Agriculture,* Washington, D.C.: U. S. Government Printing Office, 1959.

Vacu Dry Co. *Low Moisture Foods,* Emeryville, California, 1972.

West, Betty M. *Diabetic Menus, and Recipes,* Garden City, New York: Doubleday & Co., 1949.

Wigginton, Eliot, ed. *The Foxfire Books I & II,* Garden City, New York: Anchor Press/Doubleday, 1971.

Zabriskie, Bob R. *Family Storage Plan,* Salt Lake City: Bookcraft. 1969.

Index

- G -

- T -

- V -